A Practical Handbook of
Speech Coders

A Practical Handbook of
Speech Coders

Randy Goldberg
Lance Riek

CRC Press
Boca Raton London New York Washington, D.C.

Library of Congress Cataloging-in-Publication Data

Goldberg, Randy G.
    A practical handbook of speech coders / Randy G. Goldberg, Lance Riek.
        p. cm.
    ISBN 0-8493-8525-3 (alk. paper)
        1. Speech processing systems—Handbooks, manuals, etc. I. Riek, Lance. II. Title.
    TK7882.S65. G66 2000
    621.382′8—dc21
                                                  00-026994

© 2000 by CRC Press LLC
No claim to original U.S. Government works
International Standard Book Number 0-8493-8525-3
Library of Congress Card Number 00-026994
Printed in the United States of America 1 2 3 4 5 6 7 8 9 0
Printed on acid-free paper

Authors

Randy Goldberg received his bachelor's and master's degrees in 1988 from Rensselaer Polytechnic Institute. He was awarded a doctorate from Rutgers University in 1994. His background includes more than 10 years experience in speech processing, and he has authored several patents in speech coding including the Perceptual Speech Coder, the Dual Codebook Excited Linear Prediction Coder, and a fundamental patent concerning audio streaming for Internet applications. He is currently an engineering manager working in speech processing at AT&T.

Lance Riek graduated from Carnegie Mellon University in 1987 with a bachelor's degree in Electrical Engineering. He earned his Master of Engineering from Dartmouth College in 1989. He worked for six years in the Speech Processing Group of the Signal Processing Center of Technology at Sanders, a Lockheed Martin company. There, his research and development efforts focused on speech coding, speaker adaptation, and speaker and language identification. He is currently an independent engineering consultant.

To my parents,

James and Ann Riek

*For nurturing the desire to learn,
and teaching the value of work.*

Lance

To my wife, Lisa,

Randy

Acknowledgments

We would like to thank Judy Reggev, Dr. Daniel Rabinkin, and Dr. Kenneth Rosen for their feedback and suggestions. Christine Raymond was instrumental in the preparation of diagrams and overall editing, and we are grateful for her assistance. We owe a debt of gratitude to Dr. John Adcock for his significant contributions with technical revisions.

Lance Riek
Randy Goldberg

It is rare that one is fortunate enough to associate with a kind sage who is generous enough to share his lifelong learnings. During the early 1990s, I performed my Ph.D. research under the direction of Dr. James L. Flanagan. I would like to take this opportunity to thank Dr. Flanagan for all of his scholarly guidance that has had such a positive impact on my life.

Randy Goldberg

Contents

List of Figures

List of Tables

Chapter 1

Introduction

Research, product development, and new applications of speech coding have all advanced dramatically in the past decade. Research into new coding methods and enhancement of existing approaches has proceeded at a fast pace, fueled by the market demand for improved coders. Digital cellular and satellite telephony, video conferencing, voice messaging, and Internet voice communications are just a few of the prominent everyday applications that are driving the demand. The goal is higher quality speech at a lower transmission bandwidth. The need will continue to grow with the expansion of remote verbal communication.

In all modern speech coders, the inherently analog speech signal is first digitized. This sampling process transforms the analog electrical variations from the recording microphone into a sequence of numbers. The sequence is processed by an encoder to produce the coded representation. The coded representation is either transmitted to the decoder, or stored for future decoding. The decoder reconstructs an approximation of the original speech signal. As such, speech coding in general is a *lossy* compression.

In the most simple example, conventional Pulse Code Modulation (PCM) (the method used for digital telephone transmission for many years) relies upon sampling the signal and quantizing it using a sufficiently large range of numbers so that the the error in the digital approximation of the signal is not objectionable. This coding method strives for accurate representation of the time waveform.

The amount of information needed to code speech signals can be further reduced by taking advantage of the fact that speech is generated by the human vocal system. The process of simulating constraints of the human vocal system to perform speech coding is called *vocoding* (from voice coding). Efficient vocoders achieve high speech intelligibility at much lower bit rates than would be possible by coding the speech waveform directly. In the majority of vocoders, the speech signal is segmented, and each segment is considered to be the output response of the vocal tract to an input excitation signal. The excitation is modeled as a periodic pulse train, random noise, or an appropriate combination of both. For every short-time segment of speech, the excitation parameters and the parameters of the vocal tract model are determined and transmitted as the coded speech. The decoder relies on the implicit understanding of the vocal tract and excitation models to reconstruct the speech.

Some vocoders perform a frequency analysis. Manipulation of the frequency representation of the data enables easy implementation of many speech processing functions, including identification and elimination of perceptually unimportant signal components. The unimportant information can be removed, instead of wasting precious transmission data space by coding it. That saved transmission space can be reallocated to improve the speech quality of more perceptually crucial regions of the signal. Therefore, by coupling the effects of the human auditory system with those of the human vocal system, significant gains in the quality of reproduced speech can be realized for a given transmission bandwidth.

Beyond the bit rate/quality tradeoff, a practical speech coder must limit the computational complexity of the algorithm to a reasonable level for the desired application. For speech coding applications aimed at real-time or conversational communication, the overall delay must remain acceptably small. The delay is the time lag from when the speech signal was spoken at the input to when it is heard at the output. The total delay is the sum of the transmission delay of the communications system, the computational delays of the encoder and decoder, and the algorithmic delay associated with the coding method.

Speech coding can be summarized as the endeavor to reduce the transmission bandwidth (bit rate) of the coded speech through an efficient, minimal representation of the speech signal while maintaining an acceptable level of *perceived* quality of the decoded speech.

This book covers the basics of speech production, perception, and digital signal analysis techniques. These serve as building blocks to understand the various speech coding methods and their particular implementations. The presentations assume no prior knowledge of speech processing and are designed to be accessible to anyone with a technical background.

Chapter 2 provides a brief overview of speech production mechanisms and examples of speech data. This chapter introduces the concept of separating the speech signal into vocal tract and excitation components. Chapter 3 begins with sampling theory and continues with basic digital signal processing techniques that are applied in most speech analyses. Linear Prediction (LP) is explained in Chapter 4. LP modeling of the vocal tract is a primary processing step of many speech coders. Chapter 5 continues with the speech-specific processing algorithms that estimate the pitch period, or fundamental frequency, of the excitation. Accurate pitch estimation is critical to the performance of most of the newer low bit-rate systems because much of the subsequent processing depends on the pitch estimate. Human auditory processing is outlined in Chapter 6 to give a better understanding of speech perception.

Chapter 7 elaborates on scalar and vector quantization, pulse code modulation, and waveform coding. Chapter 8 discusses the evaluation of the quality of encoded/decoded speech. Chapter 9 begins the discussion of vocoders by describing several simple types. Chapter 10 continues the presentation with LP-based vocoders that employ analysis-by-synthesis to estimate the excitation signal. In Chapter 11, the current leading approaches to low bit-rate coding are outlined. These methods model the excitation as a mixture of harmonic and noise-like components. Chapter 12 explains how the perceptual considerations of Chapter 6 can be applied to improve coder performance. Appendix A lists Internet sites that contain documentation, encoded/decoded speech examples, and software implementations for several speech coding standards.

Chapter 2

Speech Production

In order to gain a complete grasp of speech coding, one must understand, and be able to utilize, the properties of human speech production and the human listening process. Knowledge of the linguistic, physiological, and acoustic levels of speech and hearing is helpful. One must also understand current technology in voice coding, information quantization, auditory processing, and the way the properties of the human auditory system have been utilized in present day acoustic coders to reduce coding bandwidth.

Speech coding can be performed much more efficiently than coding of arbitrary acoustic signals due to the fact that speech is always produced by the human vocal tract. This additional constraint defines and limits the structure of the speech signal.

This chapter begins with a discussion of what transpires when two people communicate verbally. The role of the human vocal organs in producing speech is described in the context of the type of excitation and the impact of the vocal tract. Carrying the presentation further, specific vocal configurations are shown to produce the different phonemes of a language. The chapter concludes with the concept of the source-filter model of speech production. The source-filter model forms the basis for most low bit-rate voice coders.

2.1 The Speech Chain

A helpful way of demonstrating what happens during the speech process is to describe the simple example of two people talking to each

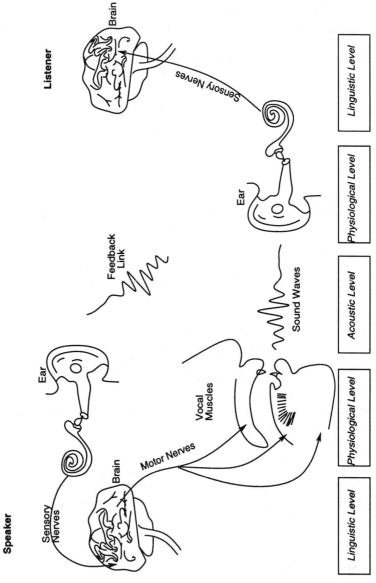

FIGURE 2.1
The speech chain (from Denes and Pinson [29]).

other; one of them, the speaker, transmits information to the other, the listener. The chain of events employed in transmitting this information will be referred to as *the speech chain* [29], and is diagrammed in Figure 2.1. The speaker first arranges his thoughts, decides what he wants to say, and puts these thoughts into a *linguistic form* by selecting the appropriate words and phrases and placing these words in the correct order as required by the grammatical structure of the language. This process is associated with activity in the speaker's brain where the appropriate instructions, in the form of impulses along motor nerves, are sent to the muscles that control the vocal organs: the tongue, the lips, the jaw, and the vocal cords. These nerve impulses cause the vocal muscles to move in such a way as to produce slight pressure changes in the surrounding air that propagate through the air in the form of a sound wave.

The sound wave propagates to the ear of the listener and activates the listener's hearing mechanism. The hearing mechanisms in the ear produce nerve impulses that travel along the acoustic nerve (a sensory nerve) to the listener's brain. When the nerve impulses arrive in the brain via the acoustic nerve, the considerable neural activity already taking place is heightened by the nerve impulses from the ear. This modification of brain activity brings about recognition and understanding of the speaker's message.

The speaker's auditory nerve supplies feedback to the brain. The brain continuously compares the quality of sounds produced with the sound qualities intended to be produced, and makes the adjustments necessary to match the results with the intended speech [29]. A lack of such feedback is partially why the hearing impaired have difficulty speaking clearly and properly.

This discussion shows how speech starts on the *linguistic level* of the speech chain in the speaker's brain through the selection of suitable words and phrases, and ends on the *linguistic level* in the listener's brain which deciphers the neural activity brought about through the acoustic nerve. Speech descends from the linguistic level to the physiological level as it is being pronounced and then into the acoustic level. The listener then brings it back to the physiological level during the hearing process and deciphers the sensations caused in this level into the linguistic level. Considering the processes that take place in each of these levels assists in understanding and developing speech coders.

2.2 Articulation

The vocal tract is the path through the human vocal organs that produce speech. The particular acoustic sound that is created is dependent on the action and position of the vocal organs. The vocal organs shape the frequency characteristics of the vibrating air traveling through the vocal tract.

The acoustic speech signal is a remarkably dynamic, complex waveform. From a signal analysis viewpoint, observing the distribution of energy across frequency for short time segments of the speech signal reveals many variations. This energy distribution across the frequency range is called the *power spectrum* or, more commonly, the *spectrum*. The energy in the spectrum can be lumped at high frequencies or low, or be evenly distributed across frequency. The fine structure of the spectrum can be random or display a definite harmonic character similar to that of musical tones. Furthermore, the variations of the spectrum over time add an additional dimension to the complexity. More than the relatively steady-state portions of the speech signal, the transitions characterize natural speech in how it sounds and, indeed, even much of the information it carries.

The many complexities of the acoustic speech signal are easier to sort and grasp when the different physiological production mechanisms are understood. By examining the vocal organs and their actions, the varying modes of the speech signal can be considered individually.

Figure 2.2 displays a simplified schematic of the primary vocal operators of the vocal tract. The diaphragm expands and contracts assisting the lungs in forcing air through the trachea, across the vocal cords and finally into the nasal and oral cavities. The air flows across the tongue, lips, and teeth and out the nostrils and the mouth. The glottis (opening formed by vocal cords or vocal folds) can allow the air from the lungs to pass relatively unimpeded or can break the flow into periodic pulses. The velum can be raised or lowered to block passage, or allow acoustic coupling, of the nasal cavity. The tongue and lips, in conjunction with the lower jaw, act to provide varying degrees of constriction at different locations. The tongue, lips, and jaw are grouped under the title *articulators*, and a particular configuration is called an articulatory position or *articulatory gesture*.

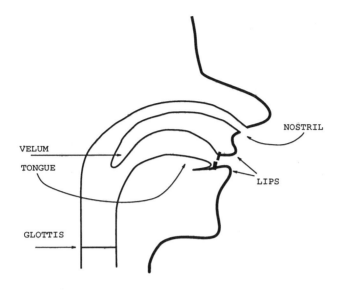

FIGURE 2.2
The primary articulators of the vocal tract.

2.2.1 Excitation

Speech sounds are produced as air is pushed from the lungs and converted into fluctuating energy. This air source and the nature of its flow are referred to as the excitation signal. It is the source of energy to excite the resonant qualities of the vocal tract. The vocal cords vibrate and form pressure pulses near the glottis, which in turn, propagate towards the oral and nasal openings. The excitation contains energy at many frequencies, and the relative strengths of these frequencies are altered as they travel through the vocal tract.

In the broadest generalization, the excitation can be considered to be *voiced* or *unvoiced*. Sounds that are created solely by the spectral shaping of the glottal pulses are called voiced sounds. All of the vowels and some consonants in the English language are voiced sounds. A sound that is pronounced without the aid of the vocal chords is called unvoiced. Unvoiced sounds are produced when air is forced through a constriction in the vocal tract and then spectrally shaped by passing through the remaining portion of the vocal tract. Sounds such as "s" and "p" are unvoiced sounds. The voiced or unvoiced character depends on the mechanism of how the excitation is produced:

1. Chopping up the steady flow of air from the lungs into quasi-periodic pulses by the vocal cords.

   - Energy is provided in this way for excitation of *voiced* sounds such as vowels.

2. Steady flow of air from the lungs with noise-like turbulence being created at some point in the vocal tract due to a constriction.

   - Energy is provided in this way for excitation of unvoiced sounds such as the sound of "s".

To restate, opening and closing of the vocal cords produces periodic, voiced excitation. A constriction on steady air flow, after the glottis, causes the noisy turbulence of unvoiced excitation.

Because the two types of excitation are produced by different mechanisms at different places in the vocal tract, it is also possible to have both present at once in a *mixed* excitation. The simultaneously periodic and noisy aspects of the sound "z" is one example. How to classify such a sound depends on the viewpoint: from a phonetic view, the sound "z" has a periodic excitation, so it is considered to be voiced. But, from the viewpoint of wanting to represent that sound in a speech coder, both the periodic and noisy attributes are present and perceptually significant, hence the *mixed* labeling. In the following phonetic discussion of speech, the sounds will be categorized as voiced or unvoiced based on the presence or absence of the periodic excitation. However, many speech sounds do have both periodic and noisy components.

Pitch

The frequency of the periodic (or more precisely, quasi-periodic) excitation is termed the *pitch*. As such, the time span between a particular point in the opening and closing of the vocal cords to that corresponding point in the next cycle is referred to as the *pitch period*. Figure 2.3 displays a time waveform for a short (40 ms) segment of a voiced sound. The x axis is the time scale, numbered in ms. The y axis is the amplitude of the recorded sound pressure. The high amplitude values mark the beginning of the pitch pulse. The first pitch period runs from near 0 ms to about 10 ms, the second from near 10 ms to about 20 ms. The spacing between the repetitions of these pulses can be discerned as approximately 10 ms. The pitch period is 10 ms, and the pitch frequency

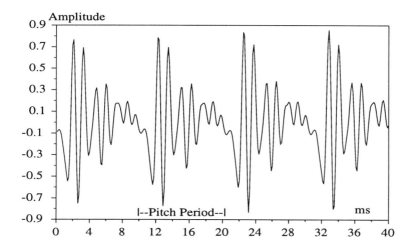

FIGURE 2.3
Time-domain waveform of a short segment of voiced speech, x-axis units in ms, y axis is relative amplitude of sound pressure.

is reciprocal of 10 ms, or 100 Hz. The pitch frequency is also referred to as the fundamental frequency.

2.2.2 Vocal Tract

The excitation is one of the two major factors affecting how speech sounds. Given the excitation as either voiced or unvoiced, the shape of the vocal tract, and how it changes shape over time, is the other primary determinant of a particular speech sound. The vocal tract has specific natural frequencies of vibration like all fluid filled tubes. These *resonant* frequencies, or *resonances*, change when the shape and position of the vocal articulators change.

The resonances of the vocal tract shape the energy distribution across the frequency range of the speech sound. These resonances produce peaks in the spectrum that are located at specific frequencies for a particular physical vocal tract shape. The resonances are referred to as *formants* and their frequency locations as the *formant frequencies*.

Figure 2.4 displays a spectrum for a short segment of voiced speech. The plot is the frequency response or frequency domain representation of

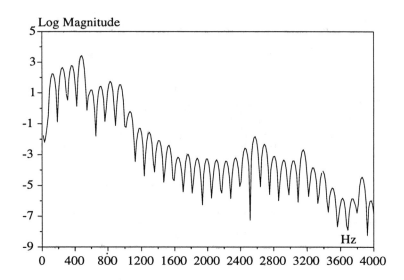

FIGURE 2.4
Log magnitude spectrum of a short segment of voiced speech,
X axis units in Hz.

the speech segment. The x axis ranges from 0 to 4000 Hz. The y axis is
the log of the magnitude of the frequency response. The narrow peaks in
the plot, regularly spaced at about 120 Hz, are the pitch harmonics. The
wider peaks in the trend of the frequency response, where a few pitch
harmonics are raised above the others, are the formant frequencies. The
first three formants are located at about 400 Hz, 900 Hz, and 2600 Hz.

The location of the formants changes significantly for different speech
sounds. The second formant, sometimes referred to as F_2, can vary as
much as 1500 Hz for a given speaker.

Manner of Articulation

In the vocal tract, the path of the airflow and the amount of constriction determine the *manner of articulation*. To produce vastly different
speech sounds, the excitation is altered by different general categories
of the vocal tract configurations. For example, vowel sounds are produced by periodic excitation, and the airflow passes through the vocal
tract mostly unrestricted. This open, but not uniform, configuration

produces the resonances associated with the formant frequencies. In a loose analogy, this is similar to the resonances produced by blowing across an open tube. Certain unvoiced sounds, called *fricatives*, have no periodic component and are the result of a steady airflow meeting some constriction. Examples of fricatives are "s" and "f."

Stop consonants, also called *stops* or *plosives*, result from the sudden release of an increased air pressure due to a complete restriction of airflow. Stops can be voiced such as sound "b" or unvoiced like the "p" sound.

Nasal consonants are produced by lowering the velum so that air can flow through the nasal cavity. At the same time, a complete constriction in the mouth prevents airflow through the lips. The most common nasal examples are "m" and "n."

Place of Articulation

The manner of articulation determines the general sound grouping, but the point of constriction, the *place of articulation*, specifies individual sounds. In other words, within the categories of sounds mentioned above, the excitation and the general arrangement of the vocal operators is the same. The different and defining attribute for a particular sound is the location of the narrowest part of the vocal tract.

Vowels sounds can be categorized by which part of the tongue produces the narrowest constriction. Examples include:

- a *front* vowel in the word "beet"

- a *mid* vowel in the word "but"

- a *back* vowel in the word "boot"

In the word "beet," the tongue actually touches the roof of the mouth just behind the teeth. In the case of "boot," the very back of the tongue, near the velum, produces the constriction.

The acoustic differences among the plosives "p," "t," and "k" are due to the different places in the vocal tract where the constrictions are made to stop the airflow before the burst.

- The constriction for "p" is closed lips.

- The constriction for "t" is the tongue at the teeth.

- The constriction for "k" is the tongue at the back of the mouth.

In short, the frequency response of the vocal tract depends upon the positions of the tongue, the lips, and other articulatory organs. The *manner of articulation* and the type of excitation (voicing) partitions English language (and most language) phonemes into broad phonetic categories. It is the *place of articulation* (point of narrowest vocal tract constriction) that enables finer discrimination of individual sounds. [128].

2.2.3 Phonemes

The qualities of the excitation and the manner and place of articulation can be considered together to classify and characterize *phonemes*. Phonemes are distinct and separable sounds that comprise the building blocks of a language. The many allowable acoustic variations of the phonemes within different contexts and by different speakers are called *allophones*. The study and classification of the speech sounds of a language is referred to as *phonetics*.

The phonemes for American English are discussed briefly for two purposes. In speech coding, it is helpful to have a grasp of speech production and the resulting range of possible acoustic variations. More importantly, an understanding of the distinct sounds of a language and how they differ is useful for coding the most basic speech information, intelligibility. When the original speech contained the phoneme /b/, but the reconstructed, coded version sounds like /g/, the message has been lost.

References [38, 137] provide more in-depth discussions of acoustic phonetics. Flanagan's reference [38] provided most of the following information. Phonemes are written with the /*/ notation. Here, the phonemes are represented as standard alphabet characters instead of phoneme symbols. This was done for simplicity and clarity. The translation to standard characters is from [137].

Vowels

Vowels are voiced speech sounds formed without significant movement of the articulators during production. The position of the tongue and amount of constriction effectively groups the vowel sounds.

Table 2.1 lists the vowels based on degree of constriction and tongue position. The words listed in the table correspond to common pronunciations; however, variations in pronunciations of these words are common. The tongue position was discussed in the previous section. The degree of

Constriction \ Position	front	mid	back
high	/i/ beet /I/ bit	/ER/ bird	/u/ boot /U/ foot
medium	/E/ bet	/UH/ but	/OW/ bought
low	/ae/ bat		/a/ father

Table 2.1 Degree of constriction and tongue positions for American English vowels.

constriction refers to how closely the tongue is to the roof of the mouth. In the phoneme /i/ ("beet"), the tongue touches the roof of the mouth. The vocal tract remains relatively wide open for the production of /ae/ ("bat").

The plots of Figures 2.5 and 2.6 display the time waveforms and log magnitude spectrums of the vowels /I/ ("bit") and /U/ ("foot"), respectively. They are presented as examples of different spectral shapes for vowels. The time waveform of /I/ displays much more high frequency characteristics than /U/. This is reflected in their spectrum plots where /I/ has a much more high-frequency energy.

It is interesting to note, for the high/back vowels, such as /U/, lip rounding is an important component of the articulatory gesture for proper production of the acoustics.

If the velum is lowered to connect the nasal passage during the vowel production, the vowel is *nasalized*. This configuration is common in French.

Fricatives

Consonants where the primary sound quality results from turbulence of the air flow, called *frication*, are grouped as *fricatives*. The frication is produced when the airflow passes a constriction in the vocal tract. Fricatives include both voiced and unvoiced phonemes.

Table 2.2 lists the fricatives. The "Constriction" column indicates the location of the constriction, which is caused by the tongue in all cases except the /f/ and /v/. In those two phonemes, the airflow is restricted by the upper teeth meeting the lower lip. The words listed in the table give common examples of the phonemes. The sound under consideration is the first sound, the leading consonant in the word, except for "vision" where it is the middle consonant sound. The term *alveolar* refers to the tongue touching the upper alveoli, or tooth sockets.

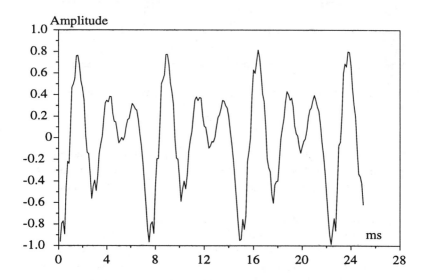

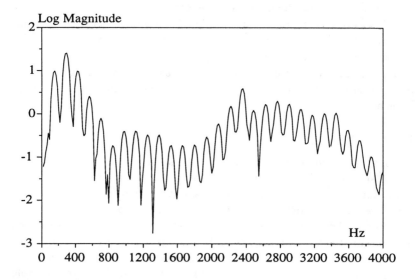

FIGURE 2.5
Time waveform and log magnitude spectrum of /I/, as in the
word "bit."

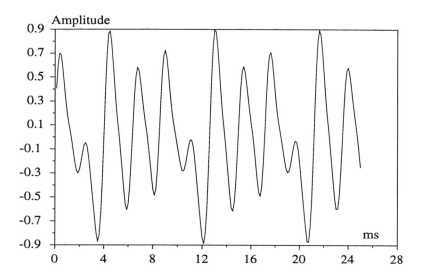

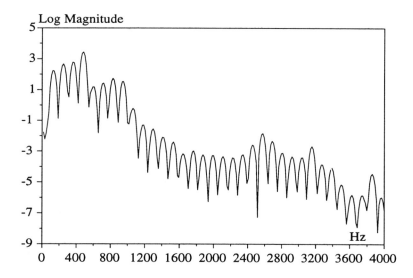

FIGURE 2.6
Time waveform and log magnitude spectrum of /U/, as in the
word "foot."

Constriction	Unvoiced	Voiced
teeth/lips	/f/ fit	/v/ vat
teeth	/THE/ thaw	/TH/ that
alveolar	/s/ sap	/z/ zip
palate	/sh/ shop	/zh/ vision (middle consonant)
glottis	/h/ help	

Table 2.2 Location of constriction and voicing for American English fricatives.

Figure 2.7 contains the time waveform and log magnitude spectrum for an example of /sh/. The sound is unvoiced, and the time waveform reflects the noise-like, random character. The spectrum has a definite shape, other than flat. The shape is imparted by the vocal tract resonances. A strong peak in the spectrum is evident at around 2800 Hz. The spectrum is indicative of the unvoiced nature; there are no regularly-spaced pitch harmonics.

Figure 2.8 displays the corresponding time waveform and log magnitude spectrum for the sound /zh/ ("vision"). It is the voiced counterpart to /sh/. The articulators are in the same position, but the excitation is periodic. The time waveform distinctly shows the noisy and periodic components of the sound. The large, regular frequency component repeats with a period of slightly less than 10 ms. On top of this, the small, irregular variations indicate the unvoiced component due to the turbulence at the constriction.

The spectrum of /zh/ shows the mixed excitation nature of the sound. The first five pitch harmonics are prominent at the low frequency end. However, across the frequency range, the fine structure of the spectrum is random, without the dominant pitch harmonics covering the entire spectrum as in the completely voiced sound of Figure 2.6. Because the articulators are in the same position as for the sound /sh/, the overall shape of the spectrum is very similar between /sh/ and /zh/. The vocal tract imparts the same shape to both the voiced excitation of /zh/ and the unvoiced of /sh/.

Stop Consonants

Stop consonants, or plosives, are formed by the release of a burst of air from a complete constriction. So, in some sense, there are two phases, the stop (complete constriction) followed by the burst (release of air).

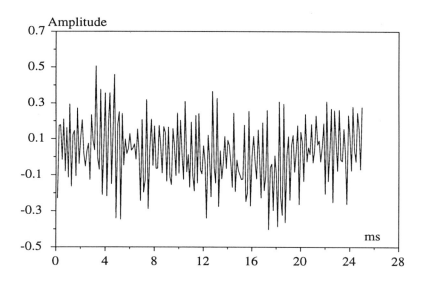

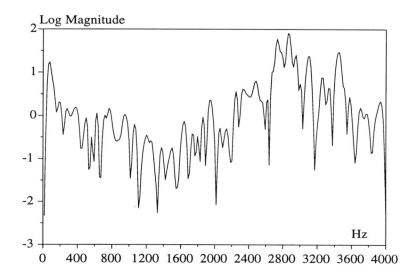

FIGURE 2.7
Time waveform and log magnitude spectrum of /sh/, as in the beginning of the word "shop."

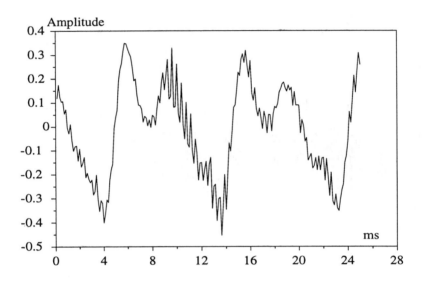

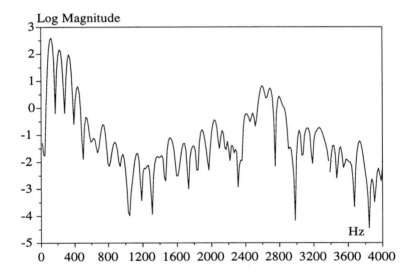

FIGURE 2.8
Time waveform and log magnitude spectrum of /zh/, as in the
middle consonant sound in the word "vision."

Constriction	Unvoiced	Voiced
lips	/p/ pat	/b/ bat
alveolar	/t/ tap	/d/ dip
back of palate	/k/ cat	/g/ good

Table 2.3 Location of constriction and voicing for American English stop consonants.

As such, they are transient sounds, short in duration. Stops can be voiced or unvoiced. The stop consonants of English are shown in Table 2.3. The constriction can be located at the lips, just behind the teeth, or at the roof of the mouth back near the velum. Table 2.3 includes common words containing the phonemes where the first sound is the stop consonant.

Figure 2.9 graphs a time waveform of the /t/ as said in context at the beginning of the word "tap." The plosive is seen primarily as one impulse, with a large negative pulse followed by a large positive pulse.

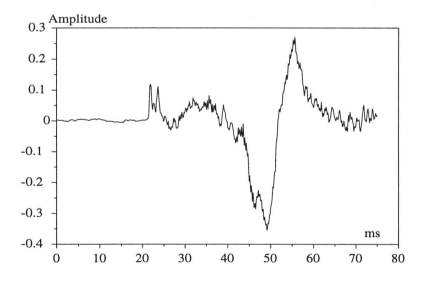

FIGURE 2.9
Time waveform of /t/, as at the beginning of the word "tap."

Because of the short, transient nature of the sound, and the articulatory gestures used to form them, stops are greatly influenced by the sounds immediately before and after. Their context can reduce them to little more than a pause (the stop) of vocalization along the trajectory from the preceding articulatory gesture to the following one. In such cases, the sound is very short in duration, of low energy, and easily confused with other stop consonants in any nonideal situation, including distortion caused by speech coding. If the stop occurs at the end of a phrase, it is often aspirated, followed by a breathy release of air.

Nasals

The defining attribute of a nasal consonant is a lowered velum which allows acoustic coupling of the nasal cavity. Nasals are voiced consonants. For nasals, the oral vocal tract is closed to airflow, and that flow is redirected out the nostrils.

Table 2.4 lists the three nasal consonants of English. Because of the closure of the oral cavity, nasals are lower in energy than most other voiced consonants. The travel of the airflow through the nasal cavity, combined with the internal acoustic coupling of the oral cavity behind the closure, results in a spectral shape different from other sounds. In short, the physical arrangement of the vocal tract produces notches in the spectrum. These are called nulls or zeros. They impact speech coding and modeling of the vocal tract for nasals.

Constriction	Voiced
lips	/m/ map
alveolar	/n/ no
back of palate	/ng/ hang (ending consonant)

Table 2.4 Location of constriction for American English nasal consonants.

Figure 2.10 plots the time waveform and spectrum for the nasal /m/. Both the time and frequency plots indicate the periodic, voiced nature of the sound. Closer examination of the spectrum reveals nulls located at approximately 900, 1700, and 3200 Hz.

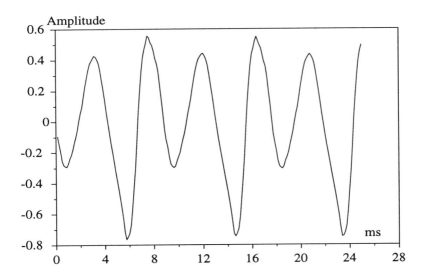

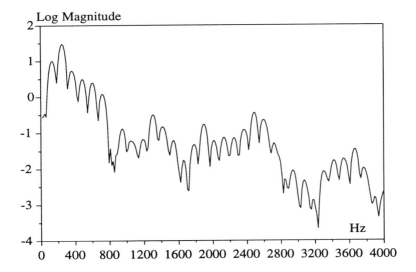

FIGURE 2.10
Time waveform and log magnitude spectrum of /m/, as in the
initial consonant of the word "map."

Affricates

Affricates are formed by the combination of a plosive followed by a fricative. As such, they are inherently a dynamic sound. The affricate of the initial consonant of the word "jam" can be considered as the combination of /d/ followed by /zh/. This is not an independent, distinct utterance of one followed by the other, but a blending of the two. The affricate of the initial sound of "chip" is the combination of /t/ and /sh/.

Semivowels

Semivowels are voiced consonants that are similar to vowels. They consist of the group /r/, /l/, /w/, and /y/. Their particular sound, and associated dynamics depend on the context. The /r/ and /l/ can be produced as a steady-state sound, but the /w/ and /y/ are strictly dynamic sounds that must involve a change of the vocal tract configuration during their production. Even though it is possible to articulate a steady-state /r/ and /l/ sound, their sound is greatly influenced by the preceding and following phonemes. This dynamic quality is brought about as the articulators move smoothly from the preceding sound, through the semivowel, and on to the following sound. Table 2.5 lists the location of constriction for English semivowels.

Because of the inherently dynamic nature of the semivowels, a single spectrum plot cannot indicate how the sound evolves over time. To display this information, a *spectrogram* is commonly used. The spectrogram is composed of a number of short-time spectrums, each one derived from further along the time signal than the last. This information is then displayed as a two-dimensional array. Each individual spectrum is a vertical slice in the array. As such, time advances along the x axis, and frequency increases along the y axis in a bottom to top manner. The values of the individual spectra are shown as colors or gray levels in the image of the array.

Constriction	Voiced
palate	/r/ run
alveolar	/l/ lap
palate	/y/ yes
lips	/w/ wet

Table 2.5 Location of constriction for semivowels.

Figure 2.11 displays a spectrogram for the nonword sounds of /u-r-i/. These were spoken continuously as though they were a word. In this image, the darker colors indicate higher energy in the spectra. Two features are most prominent in the spectrogram. The dark/light banding, most evident at low frequencies, is the pitch harmonics. The fundamental frequency can be seen as approximately 125 Hz, and the corresponding harmonics at 250, 375, 500 Hz, etc. The other feature is the groups of dark harmonics that change frequency. These are the formants, as discussed before and displayed in individual spectra. The /u/ portion of the utterance starts at the beginning and runs to near 250 ms. The /r/ segment begins at 250 ms and continues to about 500 ms, where the /i/ finishes the speech. As can be seen, these values are not exact because in continuous speech the sounds blend together.

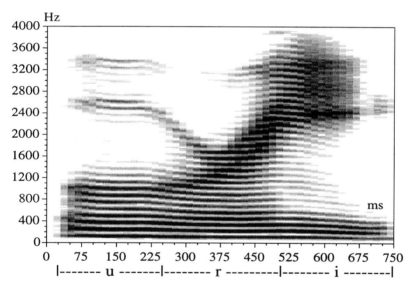

FIGURE 2.11
Spectrogram of nonword utterance /u-r-i/.

The second formant is perceptually important and changes significantly over the utterance. The second formant begins near 900 Hz for the /u/ sound, climbs steadily through 2400 Hz in the /r/, and holds near that value for the /i/. This nonword sound was chosen to illustrate the movement of the second formant for the /r/. That is, from second formant frequency of the preceding sound to that of the following one.

Figure 2.12 displays the spectrogram for the nonword utterance /i-r-u/. Here the vowels preceding and following the /r/ have switched locations. From the image, it can be seen that the track of the second formant is quite different for the /r/ than it was in Figure 2.11. In Figure 2.12, the second formant begins high, at near 2200 Hz, as it must for the vowel /i/. The vowel /i/ ranges from the beginning of the utterance to about 250 ms. At the beginning of /r/, the second formant drops sharply to 1000 Hz at the time point of 350 ms. Then it rises to 1200 Hz at 450 ms, before falling down to 900 Hz at the end of the utterance. The finish at 900 Hz is the target frequency for the /u/.

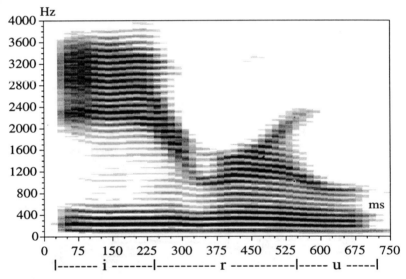

FIGURE 2.12
Spectrogram of nonword utterance /i-r-u/.

It is interesting to note the articulatory gestures required to produce /u-r-i/ and /i-r-u/. The production of /u-r-i/ is quite easy, and feels smooth and fluid as the articulatory positions for one sound blend into those for the next. This can be seen in the smooth flow of the formants in the spectrogram for /u-r-i/. Production of /i-r-u/, however, is not as smooth. The articulators have to move farther and to less similar positions to produce that sequence of sounds.

Diphthongs

Diphthongs are phonemes constructed of two vowel sounds. They are voiced, dynamic phonemes. The articulators move smoothly from the position required to produce the first vowel sound to that required for the next. This is very similar to the dynamics described for the semivowels. Diphthongs cannot be produced as a static, stationary sound.

Diphthong	Example word
/aI/	eye
/aU/	now
/oI/	boy
/oU/	no
/eI/	day
/Iu/	new

Table 2.6 The vowel combinations and word examples for American English diphthongs.

Table 2.6 lists the diphthongs for American English and common words that include them. The two letters in the phoneme symbol refer to the beginning and ending sound. The phonemes /e/ and /o/ were not included in Table 2.1 because there is some underlying ambiguity as to just what is a vowel or a diphthong. And, the exact sound and whether it should be considered a vowel or a diphthong is highly dependent on regional accent and individual speakers. The list of Table 2.6 is based on the reference of [38].

Continuous Speech Spectrogram

A spectrogram is displayed in Figure 2.13 along with the corresponding time waveform. It is presented as an example of continuous speech spectra. The phrase is "jump the lines." The affricate /j/ begins the utterance from about 50 to 100 ms. The following vowel extends from 100 to 230 ms. The lower energy nasal /m/ ranges from 230 to 280 ms. The plosive /p/ is centered at 300 ms. The voiced fricative /TH/ is very brief and located at 360 ms. Often this type of fricative is almost nonexistent in continuous speech. The vowel of the word "the" lies between 370 and 450 ms.

The lower energy consonant /l/ is between the vowels from 450 to 510 ms. The diphthong /aI/ of the word "lines" covers the time from

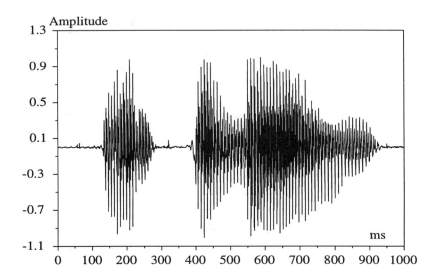

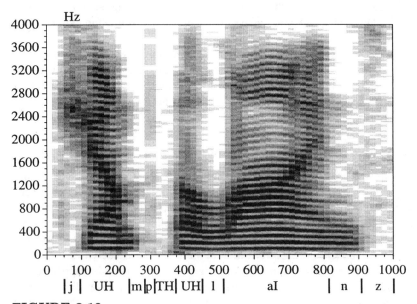

FIGURE 2.13
Time waveform and spectrogram of phrase "jump the lines."

510 to 810 ms. The nasal /n/ extends from 810 to 900 ms, and the final voiced fricative /z/ lasts from 920 to 1000 ms.

Observations to note are the smoothness of the pitch tracks and the general smoothness of the formants for voiced speech. Discontinuities in coded speech formants and pitch information due to coding errors result in degrading artifacts that are easily noticed in the reconstructed speech.

2.3 Source-Filter Model

Most voice coders (vocoders) model the vocal tract in order to simplify the analysis of the speech signal. The model is used in both the encoding and decoding processes. During encoding, the model parameters are determined to accurately represent the input speech. For decoding, the structure of the model, along with the encoded parameters, provides the guidelines for reconstructing the output speech.

A widely used speech production model is the source-filter model. The source-filter model patterns the vocal tract as a (usually linear) time-varying filter. The source energy for this filter is the excitation signal. The different ways of coding this excitation signal are generally what separates these source-filter speech coders from one another.

The source-filter model results from considering the excitation and vocal tract as separable components in the production of speech. The excitation is produced at some point in the vocal tract, and then the excitation is spectrally shaped (or filtered) by the rest of the vocal tract.

The Vocal Tract

The throat, nose, tongue, and mouth form a resonating air-filled cavity that predominantly dictates the sound produced by the human vocal system. The resonant frequencies of this tube are called *formant frequencies*. Different configurations of the vocal tract result in different formant frequencies. The formant frequencies are one of the two major factors that dictate which phoneme will be produced by the vocal tract. The other major factor is the excitation of the vocal tract.

Excitation

For voiced speech, a periodic waveform provides the excitation to the vocal tract. The periodic waveform results from the glottal pulses created by the rapid opening and closing of the vocal cords. A simple and widely used model for unvoiced speech is shaped white noise. White noise is random and has a flat spectral shape where all frequencies have equal power. The white noise is assumed to be generated when air passes through a constriction. Some sounds such as /z/ are produced by both exciting the vocal tract with a periodic excitation and by forcing air through a constriction in the vocal tract. This is called *mixed excitation*. One of the challenges in speech coding is to be able to accurately represent sounds that are voiced, unvoiced, or mixed.

General Source-Filter Model

The diagram of Figure 2.14 illustrates the flow of signals and information for a generalized source-filter model. The pitch information is usually contained in a pitch period value. The values change over time, and are estimated and updated along with the changing speech signal. Based on the pitch period, the "Periodic Excitation" block produces a pulse waveform that represents the glottal pulses. The "Noise Excitation" block outputs a noisy sequence with a flat spectral response. The two excitations are input to the mixing decision. Time varying information about the voicing of the speech is the other input. Based on the level of voicing in the original speech, the "Mixing Decision" block combines the periodic and noisy excitations in appropriate amounts to produce the excitation signal.

A classic version of the two-state, source-filter model incorporates a hard voiced/unvoiced decision for each segment of speech. In that case, the "Mixing Decision" functions as a switch, and the excitation is entirely voiced or unvoiced, depending on the classification.

The vocal tract information is fed into the "Vocal Tract" box to produce a vocal tract filter. The filter shapes the spectrum of the excitation to that of the original speech. In practice, the vocal tract information can be represented by several methods, including a *linear predictor* and *Fourier magnitudes*. These methods, along with representations of the excitation, form the central topic for most of the remainder of the book. The excitation is filtered by the vocal tract model to produce the synthesized speech. The goal is to have the synthesized speech sound, perceptually, to the human ear, as close to the original as possible.

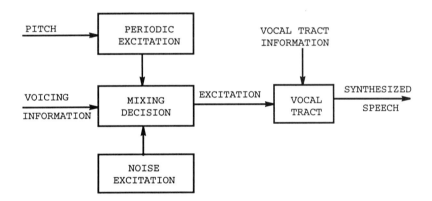

FIGURE 2.14
General source-filter model.

Chapter 3

Speech Analysis Techniques

This chapter contains a brief introduction to the signal processing concepts that are applied to speech coding. For a more thorough examination of these concepts please refer to other signal and speech processing books such as *Introduction to Signal Processing* [127], *Discrete Time Signal Processing* [126], and *Digital Processing of Speech Signals* [137].

The chapter begins with a description of sampling the analog speech waveform to produce a discretely sampled version. The concept of the input/output relations of linear systems is presented next. Frequency domain transforms are basic operations for most all types of speech processing. The general z-transform introduces the idea, followed by the Fourier transform (the z-transform evaluated on the unit circle in the z-domain). The discrete Fourier transform (DFT) is the discretely sampled version of the Fourier transform. By reorganizing the structure of the DFT, the fast Fourier transform (FFT) produces the same transform result while significantly reducing the computational complexity. A discussion of the effects of windowing of data segments completes the chapter.

3.1 Sampling the Speech Waveform

Speech signals are analog in nature because they originate as sound pressure waves. After transduction by a microphone into an electrical signal, the speech signal is still analog. However, all speech coding algorithms rely on computer processing of discretely sampled versions of the speech. To accurately represent the original signal with discrete samples requires a few guidelines.

If $s_{analog}(t)$ represents the analog speech signal, the sampled signal can be expressed as:

$$s(n) = s_{analog}(nT) \qquad\qquad (3.1)$$

where n takes on integer values, and T is the time between samples, known as the sampling period.

Though it might not seem likely, if $s_{analog}(t)$ is bandlimited (no frequency components higher than a known limit), and is sampled fast enough (T small enough), $s(n)$ provides a complete and unique representation of $s(t)$. The keys to the statement are "bandlimited" and "fast enough." In this case, fast enough is twice the highest frequency component:

$$T \leq 1/2F_{max} \qquad\qquad (3.2)$$

and is referred to as the *Nyquist rate*.

This can be illustrated in the plots of Figure 3.1. In the first plot, $s(t)$ is sampled more than twice as fast as the highest frequency. For simplicity, $s(t)$ is shown to have one dominant frequency component, but the discussion holds true for any bandlimited signal. In the second plot, $s(t)$ is sampled at two samples for each period of the major frequency component which is just barely enough to represent the signal. In the third plot, the signal is undersampled so that an ambiguity results. When considering only the discrete samples, the high frequency solid line appears identical to the low frequency dashed line due to undersampling. This misrepresentation of frequencies is known as *aliasing*. A frequency greater than 1/2 the sampling frequency in the original signal has been aliased as the lower frequency of the dotted, sampled waveform.

The aliasing can be eliminated by bandlimiting the speech before it is sampled. Speech is naturally bandlimited to have the vast majority of its energy below 7 kHz. But, to sample speech at rates below 14 kHz, or to remove the small amount of energy above that range, a lowpass filter is applied before sampling. In practice, speech is often lowpass filtered before the sampling to slightly less than 4 kHz and then sampled at 8 kHz. This is referred to as *narrowband speech*, and is the common input for most of the coders discussed in future chapters. The 4 kHz bandwidth preserves good intelligibility, speaker identity, and naturalness. However, for higher quality sound, speech can be sampled and coded at higher sampling (and coding) rates.

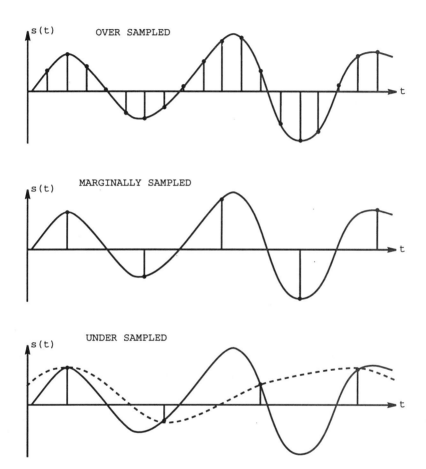

FIGURE 3.1
Illustration of sampling rate relative to the Nyquist rate.

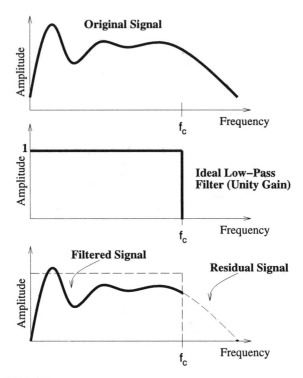

FIGURE 3.2
Ideal lowpass filter: frequency domain representation.

The goal with the lowpass filtering before sampling is to remove all of the frequency components that have frequencies greater than half the sampling rate. Figure 3.2 displays the frequency representation of an ideal lowpass anti-aliasing filter. In this case, the cutoff frequency, f_c, of the filter is less than or equal to $F_s/2$, one half the sampling frequency.

3.2   Systems and Filtering

The lowpass filter discussed in the previous section can be considered as a system. The system operates on an input signal to produce an output signal, where the output signal is changed in some desirable fashion. Figure 3.3 displays a block diagram for a filter. The input

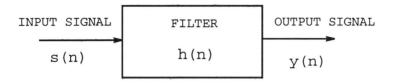

FIGURE 3.3
Discrete time filter.

signal, $s(n)$, is altered by the filter to produce the output, $y(n)$.

A system is a *linear system* if scaled and added input sequences yield corresponding scaled and added output sequences. In particular, if the input sequence $s(n)$ is expressed as:

$$s(n) = as_1(n) + bs_2(n) \tag{3.3}$$

then, $y(n)$ must be:

$$y(n) = ay_1(n) + by_2(n) \tag{3.4}$$

for $h(n)$ to be a linear system.

Filtering is a basic digital signal processing operation often used in speech coding. Filtering is the mathematical operation of the *convolution* of a digital filter with an input sequence to produce an output sequence. The convolution sum is defined as:

$$y(n) = s(n) * h(n) = \sum_{k=-\infty}^{\infty} s(k)h(n-k) \tag{3.5}$$

In Equation 3.5, the output sequence $y(n)$ is the result of passing the input sequence, $s(n)$, through the digital filter, $h(n)$.

The filter can be specified as a time-domain sequence by its *impulse response*. The impulse response of a digital filter, $h(n)$, is the sequence that results as the output from a unit sample $\delta(n)$ input. The unit sample sequence is defined as:

$$\delta(n) = \begin{cases} 1 \text{ for } n = 0 \\ 0 \text{ for } n \neq 0 \end{cases} \tag{3.6}$$

3.3   Z-Transform

Various aspects of speech coding are easier to analyze and understand with a frequency domain representation of a signal or system. The z-transform is the discretely sampled analogy to the Laplace transform for continuous signals. The z-transform provides a useful representation to analyze the spectral shaping qualities of a pole and/or zero system, and as the more general expression of the Fourier transform.

For a discrete signal $s(n)$, the z-transform is defined as:

$$S(z) = \sum_{n=-\infty}^{\infty} s(n)z^{-n} \tag{3.7}$$

The condition for convergence of the infinite series is:

$$\sum_{n=-\infty}^{\infty} |s(n)||z^{-n}| < \infty \tag{3.8}$$

If $s(n)$ is of finite length, Equation 3.8 will converge for at least all nonzero, noninfinite values of z.

The inverse transform for deriving $s(n)$ from $S(z)$ is:

$$s(n) = \frac{1}{2\pi j} \oint S(z)z^{n-1}dz \tag{3.9}$$

where the contour integral is evaluated on a closed contour, within the region of convergence for z and enclosing the origin.

The expression for a single pole in the z-plane is a useful example. A pole is a value of z for which $S(z)$ is infinite. (Conversely, a zero is a value for which $S(z)$ is zero.) For the decaying exponential sequence:

$$h(n) = \begin{cases} a^n & \text{for } n \geq 0 \ |a| < 1 \\ 0 & \text{for } n < 0 \end{cases} \tag{3.10}$$

the transform is:

$$H(z) = \sum_{n=0}^{\infty} a^n z^{-n} \tag{3.11}$$

and it converges to:

$$H(z) = \frac{1}{1 - az^{-1}} \tag{3.12}$$

Property	Time Domain	Z Domain
Linearity	$as(n) + bh(n)$	$aS(Z) + bH(z)$
Shift	$s(n + N)$	$z^N S(z)$
Multiply by a^n	$a^n s(n)$	$S(a^{-1}z)$
Convolution	$s(n) * h(n)$	$S(z)H(z)$

Table 3.1   Theorems of z-transforms.

for $|a| < |z|$. The frequency shaping attributes of multiple poles in the z-plane form the basis for linear prediction (LP) modeling of the speech spectrum that will be discussed in Chapter 4, and will be explained for the simple case of one pole in the next section.

Table 3.1 displays several important z-transform theorems. Most notable are:

- Linear combinations of signals in one domain correspond to linear combinations in the alternate domain.

- Convolution in the time domain corresponds to multiplication in the z-domain.

- A shift in the time domain corresponds to multiplication by z raised to power of the length of the shift (in samples).

The z-transform of the system of Equation 3.5 results in:

$$Y(z) = S(z)H(z) \qquad (3.13)$$

so that convolution in the time domain transforms to multiplication in the frequency domain. $Y(z)$, $S(z)$, and $H(z)$ are the z-transforms of $y(n)$, $s(n)$, and $h(n)$.

If the impulse response of a filter, $h(n)$, is of a finite length (nonzero for limited range of n, zero outside that range), then $h(n)$ is known as a finite impulse response (FIR) filter or system. Conversely, if $h(n)$ has an infinite duration of nonzero values, it is termed an infinite impulse response (IIR) filter. The single pole system of Equations 3.10 and 3.12 is an IIR system – the values of $h(n)$ are nonzero for all positive n. In general, all-pole systems are IIR systems.

If $h(n)$ is of finite length (FIR), the function $H(z)$ is a polynomial in the variable of z^{-1}. $H(z)$ will have zeros, but no poles for all nonzero values of z.

3.4   Fourier Transform

The Fourier transform represents a signal in terms of complex expo-
nentials (or sinusoids, because $e^{-j\omega n} = \cos(\omega n) - j\sin(\omega n)$). As such,
the frequency representation of a signal through the Fourier transform
facilitates some processing and signal visualizations that are inherently
frequency oriented.

The discrete-time Fourier transform pair defined by the forward transform:

$$S(\omega) = \sum_{n=-\infty}^{\infty} s(n)e^{-j\omega n} \qquad (3.14)$$

and the inverse transform:

$$s(n) = \frac{1}{2\pi} \int_{-\pi}^{\pi} S(\omega)e^{j\omega n}d\omega \qquad (3.15)$$

is a mathematical link between the frequency representation and the
time representation of a time sequence $s(n)$. $S(\omega)$ is the *frequency response* of $s(n)$. It can be seen that Equation 3.14 is the same as Equation
3.7 where $z = e^{j\omega}$; it is the evaluation of the z-transform along the unit
circle in the z-plane. For all cases, $S(\omega)$ is a periodic signal with a period
of 2π.

The relations for the z-transform of Table 3.1 hold for their corresponding Fourier transforms analogies.

In general, $S(\omega)$ is a complex signal. For a real time-domain signal
$s(n)$, the real part and the magnitude of $S(\omega)$ are even, and the imaginary part and phase are odd.

$$Re[S(\omega)] = Re[S(-\omega)] \qquad \text{and} \qquad |S(\omega)| = |S(-\omega)| \qquad (3.16)$$

$$Im[S(\omega)] = -Im[S(-\omega)] \qquad \text{and} \qquad \angle S(\omega) = -\angle S(-\omega) \qquad (3.17)$$

As such, the Fourier transform for a real signal is specified completely
by the range $0 \leq \omega \leq \pi$. For the range $-\pi \leq \omega \leq 0$, the magnitude and
real part are "flipped" left for right; and the phase and imaginary part
are flipped and inverted.

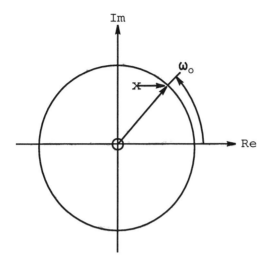

FIGURE 3.4
Pole/zero configuration in the z-plane for a single pole system, pole at "x," zero at origin.

For a filter with impulse response, $h(n)$, the Fourier transform of $h(n)$ is the frequency response, $H(\omega)$. Convolution in the time domain transforms to multiplication in the frequency domain such that:

$$Y(\omega) = S(\omega)H(\omega) \tag{3.18}$$

is the frequency domain representation of Equation 3.5 where $Y(\omega)$, $S(\omega)$, and $H(\omega)$ are the Fourier transforms of $y(n)$, $s(n)$, and $h(n)$.

The frequency response of the single pole example of Equation 3.12 can be visualized and evaluated graphically. Equation 3.12 can be multiplied by z/z to yield the equivalent:

$$S(z) = \frac{z}{z - a} \tag{3.19}$$

showing that the expression has a pole at $z = a$ and a zero at $z = 0$. The frequency response (Fourier transform) is found by evaluating the expression on the unit circle ($z = e^{j\omega}$). Figure 3.4 shows the location of the pole and zero in the complex z-plane. As noted, $|a| < |z|$ and z must be able to be evaluated on the unit circle, so $|a| < 1$.

The magnitude of the frequency response is evaluated at frequency ω_o. Here, ω_o is the angle swept from the real axis, counterclockwise, to the point on the unit circle marked as ω_o. The magnitude is the length of the vector from the zero at the origin to frequency ω_o on the unit circle, divided by the length of the vector from the pole. Imagine sweeping the location of ω_o around the unit circle. The length of the zero vector remains the same at 1. The length of the pole vector will be quite small near the location of the pole and nearly 2 on the opposite side. So, the magnitude of the frequency response of the single pole system will show a sharp peak at the frequency ω near the pole location. This peak will be sharper and higher in magnitude as the pole is moved closer to the unit circle.

3.5   Discrete Fourier Transform

Because most speech signals are not known over all time, the Fourier transform does not exist without modification of the speech signal (and the transform). The discrete Fourier transform (DFT) is a much more usable frequency transformation of a speech waveform. The DFT is a Fourier representation of a sequence of samples of limited length. Instead of being a continuous function of frequency as the FT, the DFT is a sequence of samples. The samples of the DFT are equally spaced along the frequency axis of the FT.

The DFT is defined as:

$$S(k) = \sum_{n=0}^{N-1} s(n)e^{-j\frac{2\pi}{N}kn} \tag{3.20}$$

where N is the length of the segment.

In this formulation, $s(n)$ is considered to be periodic with a period of N. That is, $s(n)$ repeats the finite sequence for all n. Also, the DFT can be thought of as sampling the Fourier transform at N evenly spaced points on frequency axis (the unit circle in the z-plane).

The inverse DFT is given by:

$$s(n) = \frac{1}{N} \sum_{n=0}^{N-1} S(K)e^{j\frac{2\pi}{N}kn} \tag{3.21}$$

It can be seen that the inverse transform is the same as the forward transform, except for a scale factor $\frac{1}{N}$ and sign change of the exponential argument.

In addition to the problem of not knowing the speech signal over all time, speech is highly *nonstationary* – the statistics of the signal change over time. Indeed, the very information-carrying nature of the signal is responsible for these changes. When the properties of a signal are invariant to a shift in the time index, the signal is referred to as a *stationary* signal. When one listens to different time instances of a spoken sentence, completely different phonemes are heard. Different frequency components exist in the different time instances of the spoken sentence. Although a speech signal is not stationary, it is *quasi-stationary* in that a small segment of speech (20ms or less) "pretty much" has the properties of a stationary signal.

For this reason, the Fourier transform of small segments of speech is extremely valuable in speech signal processing. What is usually desired is a running spectrum with time as an independent variable in which the spectral computation is made on windowed, and weighted past values of the signal [41]. A specified time interval (e.g., 20 ms) is used for the segment, and the segment is weighted accordingly (see Section 3.6). The truncated weighted segment of speech is Fourier transformed. The resulting frequency parameters are associated with the time segment of speech corresponding to the center of the analysis interval. Consequently, the transform is adapted to:

$$S(k) = \sum_{n=0}^{N-1} s(n)w(n)e^{-j\frac{2\pi}{N}kn} \tag{3.22}$$

where $w(n)$ is the windowing function and N is the length of the window.

The usual definition for the DFT of a segment of speech, without the window, is equivalent to a rectangular window. In that case, the window $w(n) = 1$ for $0 \leq n \leq N$ and 0 outside that range.

3.5.1 Fast Fourier Transform

The fast Fourier transform (FFT) is a group of methods that rearrange the calculations in the DFT to allow significant computational savings. Direct computation of the DFT requires a number of multiplies on the order of N^2, while the FFT reduces that number to the order of $N \log N$.

The process uses the symmetry and periodicity of the exponential factor to reduce the computations [126]:

$$e^{-j\frac{2\pi}{N}k(N-n)} = \left(e^{-j\frac{2\pi}{N}kn}\right)* \tag{3.23}$$

$$e^{-j\frac{2\pi}{N}kn} = e^{-j\frac{2\pi}{N}k(n+N)} = e^{-j\frac{2\pi}{N}n(k+N)} \tag{3.24}$$

where $(.)*$ denotes the complex conjugate operation.

The FFT works by recursively decomposing the N-point DFT into smaller DFTs. The implementation is most common for powers of 2 because of the convenience in fitting the recursive structure. However, the FFT method can be applied to any sequence length that is a product of smaller integer factors.

The *decimation in time* FFT algorithm begins by separating the N length sequence $s(n)$ into two $N/2$ length sequences in the DFT computation:

$$S(k) = \sum_{neven} s(n)e^{-j\frac{2\pi}{N}kn} + \sum_{nodd} s(n)e^{-j\frac{2\pi}{N}kn} \tag{3.25}$$

and by substituting $n = 2m$ for even n, and $n = 2m + 1$ for odd n, Equation 3.25 can be expressed as:

$$S(k) = \sum_{m=0}^{(N/2)-1} s(2m)e^{-j\frac{2\pi}{N}k2m} + \sum_{m=0}^{(N/2)-1} s(2m+1)e^{-j\frac{2\pi}{N}k(2m+1)} \tag{3.26}$$

which is rewritten as:

$$S(k) = \sum_{m=0}^{(N/2)-1} s(2m)e^{-j\frac{2\pi}{N/2}km} + e^{-j\frac{2\pi}{N}k} \sum_{m=0}^{(N/2)-1} s(2m+1)e^{-j\frac{2\pi}{N/2}km}$$
$$\tag{3.27}$$

where both of the sums are now arranged into an $N/2$-point DFT. As such, the N-point DFT has been decomposed into the sum of two $N/2$-point DFTs, with one multiplied by $e^{-j\frac{2\pi}{N}k}$.

At the next stage, the two $N/2$-point DFTs are decomposed into four $N/4$-point DFTs. This process is repeated at each stage until the whole DFT is decomposed into $N/2$ 2-point DFTs.

For details on the implementation of the FFT, refer to a signal processing text such as [126].

3.6 Windowing Signal Segments

The window function, $w(n)$, introduced in the previous section, serves not only to select the correct segment of speech for processing, but also to weight the speech samples of $s(n)$. The selected segment of speech is referred to as the speech *frame*. The shape of the window affects the frequency representation, $S(k)$, by the frequency response of the window itself. As mentioned in this chapter, convolution in the time domain is multiplication in the frequency domain. Conversely, multiplication in the time domain corresponds to convolution in the frequency domain. The multiplication of a time-domain speech sequence $s(n)$ with a time-domain window $w(n)$ is the same as the convolution of $S(k)$ and $W(k)$ in the frequency domain. So, the impact of a window shape can be analyzed by examining its DFT.

Figure 3.5 displays the time-domain shapes for two windows of length 300 samples. The Hamming window is the dotted line. The Hamming and Hanning are both raised cosine functions with similar frequency characteristics. The popular Hamming (raised at the edges) features good attenuation of the first few sidelobes and a nearly flat response for the higher frequency sidelobes. The first few sidelobes of the Hanning are higher in amplitude, but the higher frequency sidelobes continue to roll off to negligible low values.

The Hamming window is given by:

$$w(n) = 0.54 - 0.46 \cos \frac{2\pi n}{N}, \qquad \text{for } 0 \le n \le N - 1 \qquad (3.28)$$

and the Hanning by:

$$w(n) = 0.5 - 0.5 \cos \frac{2\pi n}{N}, \qquad \text{for } 0 \le n \le N - 1 \qquad (3.29)$$

The frequency responses of the Hamming, Hanning, and rectangular windows are shown in Figure 3.6. As can be seen, the main lobe of the rectangular window is about half as wide as the Hamming or Hanning. The side lobes are much lower for the Hamming and Hanning than the rectangular. The first side lobe of the Hanning is approximately 20 dB higher than the Hamming, but the Hanning sidelobes rapidly decrease to very low levels.

Selecting the window shape, and its resulting frequency response, is a tradeoff between a narrow main lobe in the frequency domain and low

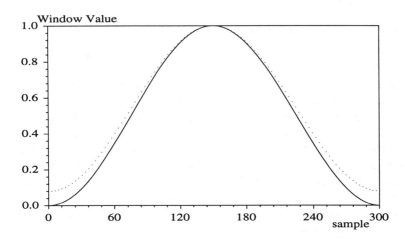

FIGURE 3.5
Window shapes for Hamming (dotted) and Hanning windows.

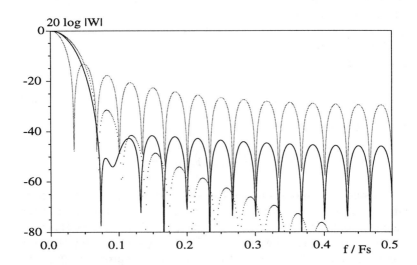

FIGURE 3.6
Frequency responses for Rectangular (gray), Hamming (black),
and Hanning (dotted) windows.

sidelobes. A narrow main lobe improves frequency resolution so that in the resulting DFT magnitude, closely spaced, narrow components are separated. A narrow main lobe, as in the rectangular window, comes at the expense of high side lobes. These side lobes add a noisy appearance to the DFT magnitude due to interference from adjacent harmonics and make it more difficult to discriminate low magnitude components.

For the Hamming window, the approximate bandwidth of the main lobe is:

$$BW = \frac{2F_s}{N} \tag{3.30}$$

where N is the length of the window in number of samples.

It is easier to interpret the impact of window selection on the Fourier transform of a speech segment with an example. Figure 3.7 displays a segment of voiced speech and the corresponding DFT log magnitude. In this case, no explicit window was used, except a rectangular window to excise this segment of data from the longer speech signal. For comparison, the same segment of speech was windowed with a Hamming window and transformed with the DFT for the plots of Figure 3.8.

The Hamming-windowed, time-domain segment displays the influence of the center-weighted, symmetrically decaying ends of the window shape. The rectangular-windowed DFT magnitude of Figure 3.7 displays high resolution in the locations of the pitch harmonics in the lower frequencies around 100 to 1000 Hz. However, in the mid frequencies of 1500 to 2500 Hz, the structure appears noisy, making it difficult to distinguish the pitch harmonics. This is caused by *spectral leakage* where the energy associated with one pitch harmonic obscures neighboring harmonics.

Conversely, in the plot of the DFT magnitude of the Hamming-windowed segment of Figure 3.8, the peaks of the pitch harmonics are wider. But, the harmonics are clearly represented in the entire frequency range, including the 1500 to 2500 Hz range. The lower sidelobes of the Hamming window prevent spectral leakage.

The spectrum plots and spectrogram presented in Chapter 2 were computed by this method of taking the log magnitude of the DFT of the Hamming-windowed speech segment. For the spectrogram, the window slides along the speech signal in increments smaller than the window. These spectra are compiled and displayed with each spectrum representing a vertical column in the spectrogram image.

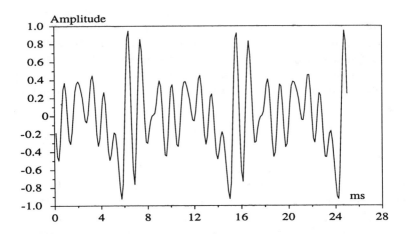

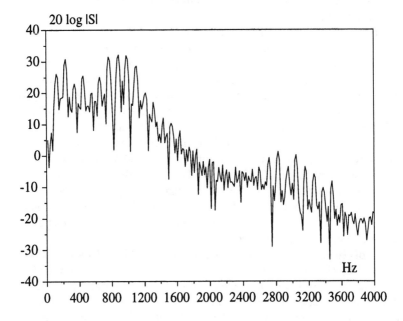

FIGURE 3.7
**Time-domain speech segment and DFT log magnitude with
rectangular window.**

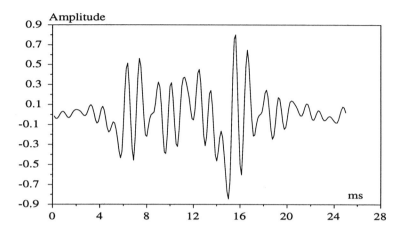

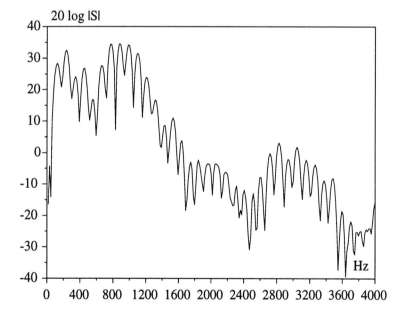

FIGURE 3.8
Windowed time-domain speech segment and DFT log magnitude with Hamming window.

Chapter 4

Linear Prediction Vocal Tract Modeling

Linear Prediction (LP) is a widely used and successful method that represents the frequency shaping attributes of the vocal tract in the source-filter model of Section 2.3. For speech coding, the LP analysis characterizes the shape of the spectrum of a short segment of speech with a small number of parameters for efficient coding. Linear prediction, also frequently referred to as Linear Predictive Coding (LPC), predicts a time-domain speech sample based on a linearly weighted combination of previous samples. LP analysis can be viewed simply as a method to remove the redundancy in the short-term correlation of adjacent samples. However, additional insight can be gained by presenting the LP formulation in the context of lossless tube modeling of the vocal tract.

This chapter presents a brief overview of the the lossless tube model and methods to estimate the LP parameters. Different, equivalent representations of the parameters are discussed along with the transformations between the parameter sets. Reference [137] discusses the lossless tube model in great detail.

4.1 Sound Propagation in the Vocal Tract

Sound waves are pressure variations that propagate through air (or any other medium) by the vibrations of the air particles. Modeling these waves and their propagation through the vocal tract provides a framework for characterizing how the vocal tract shapes the frequency content of the excitation signal.

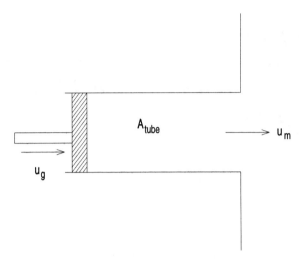

FIGURE 4.1
Diagram of uniform lossless tube model.

Modeling the vocal tract as a uniform lossless tube with constant cross-sectional area is a simple but useful way to understand speech production. A diagram of this model is shown in Figure 4.1. In the figure, u_g and u_m represent the volume velocity flow at the glottis and mouth, respectively; and A_{tube} is the constant cross-sectional area of the tube.

A system of partial differential equations describes the changes in pressure and volume velocity over time and position along the tube. Assuming ideal conditions (no losses due to viscosity or thermal conduction and no variations in air pressure at the open end of the tube), Portnoff's wave equations [135, 10] characterize this system as:

$$-\frac{\partial p}{\partial x} = \rho \frac{\partial (u/A)}{\partial t} \tag{4.1}$$

and

$$-\frac{\partial u}{\partial x} = \frac{1}{\rho c^2}\frac{\partial (pA)}{\partial t} + \frac{\partial A}{\partial t} \tag{4.2}$$

where:

$x =$ location inside the tube

$t =$ time

$p(x,t)$ = sound pressure at location x and time t

$u(x,t)$ = volume velocity flow at location x and time t

ρ = density of air inside the tube

c = velocity of sound

$A(x,t)$ = cross-sectional area of the tube at location x and time t

Because $A(x,t)$ is a constant A, in this example, the wave equations can be simplified for a uniform lossless tube:

$$-\frac{\partial p}{\partial x} = \frac{\rho}{A}\frac{\partial u}{\partial t} \tag{4.3}$$

and

$$-\frac{\partial u}{\partial x} = \frac{A}{\rho c^2}\frac{\partial p}{\partial t} \tag{4.4}$$

resulting in two equations with two unknowns that are integrated with respect to time to yield the following volume velocity and pressure definitions:

$$u(x,t) = u_1\left(t - \frac{x}{c}\right) - u_2\left(t + \frac{x}{c}\right) \tag{4.5}$$

and

$$p(x,t) = \frac{\rho c}{A}\left(u_1\left(t - \frac{x}{c}\right) + u_2\left(t + \frac{x}{c}\right)\right) \tag{4.6}$$

Further examination of these formulas reveals that u_1 is a wave propagating towards the open end of the tube, while u_2 propagates toward the closed end. Also note that both the sound pressure and volume can be described by scaled addition/subtraction (superposition) of these waves.

This simple model of the vocal tract has the same properties of a simple electrical system. Comparing the wave equations of the lossless tube system to the current $i(x,t)$ and voltage $v(x,t)$ equation of a uniform lossless transmission line:

$$-\frac{\partial v}{\partial x} = L\frac{\partial i}{\partial t} \tag{4.7}$$

and

$$-\frac{\partial i}{\partial x} = C\frac{\partial v}{\partial t} \tag{4.8}$$

Equations 4.3 and 4.4 are the same as 4.7 and 4.8 with the variable substitutions shown in Table 4.1.

The frequency response of a system of this type is well known, and finding the frequency response of the lossless tube system requires only

Electrical System	Acoustic System
L (inductance)	$\frac{p}{A}$
C (capacitance)	$\frac{A}{\rho c^2}$
v (voltage)	p
i (current)	u

Table 4.1   Analogy between electrical and acoustic quantities.

the scaling shown in Table 4.1. The system has an infinite number of poles on the $j\omega$ axis corresponding to the tube resonant frequencies of $\frac{c}{4l} \pm \frac{nc}{2l}$, where $n = 0, 1, ..., \infty$. These resonances are plotted in Figure 4.2 for a limited frequency range.

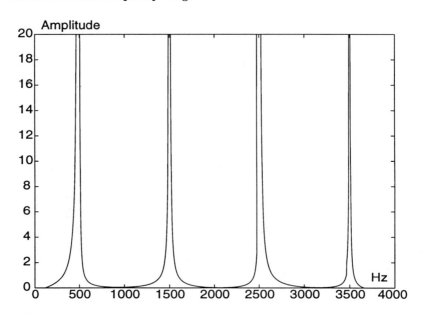

FIGURE 4.2
Frequency response of a single lossless tube system.

The frequency response of the lossless tube system is not dependent on the source, just as the impulse response of an electrical system is not dependent on its input. The resonant frequencies of the vocal tract are called *formant frequencies*. If the tube is 17.5 cm long, and 35,000 cm/sec is used as c (the speed of sound), then the formant frequencies of this system are $\frac{35,000 \ cm/sec}{4(17.5 \ cm)} \pm (n) \frac{35,000n \ cm/sec}{2(17.5 \ )} = 500 \ \text{Hz} \pm (n) \ 1000 \ \text{Hz}$ [137]. In an actual vocal tract, which is not uniform in area and is not lossless, formant frequencies are generally not as evenly spaced. A human vocal system also changes over time as the person articulates sounds. Therefore, the formant frequencies also change over time.

4.1.1 Multiple-Tube Model

In a physical vocal tract, the cross-sectional area varies based on position along the tract and over time. These variations create different speech sounds with the same excitation. To better model the varying cross-sectional area of the vocal tract, the single lossless tube can be extended to many lossless tubes concatenated to one another as depicted in Figure 4.3.

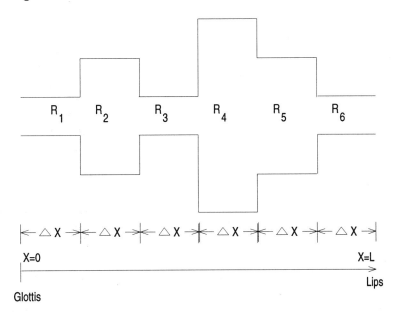

FIGURE 4.3
Multiple concatenated tube model.

The vocal tract is excited at x=0, which is either at the glottis (as depicted in Figure 4.3) or at some constriction in the vocal tract. The excitation propagates through the series of tubes with some of the energy being reflected at each junction and some energy being propagated. The

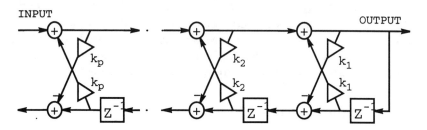

FIGURE 4.4
Lattice filter realization of multiple-tube model.

reflection coefficients signify how much energy is reflected and how much is passed. These reflections cause spectral shaping of the excitation. This spectral shaping acts as a digital filter with the order of the system equal to the number of tube boundaries.

The digital filter can be realized with a lattice structure, where the reflection coefficients are used as weights in the structure. Figure 4.4 displays the lattice filter structure. The k_i is the reflection coefficient of the ith stage of the filter. The flow of the signals suggests the forward and backward wave propagation as mentioned previously. The input is the excitation, and the output is the filtered excitation, that is, the output speech. There are p stages corresponding to p tube sections. The time delay for each stage in the concatenated tube model is $\Delta x/c$ where c is the speed of sound.

The lattice structure can be rearranged into the direct form of the standard all-pole filter model of Figure 4.5. In this form, each tap, or *predictor coefficient*, of the digital filter delays the signal by a single time unit and propagates a portion of the sample value. There is a direct conversion between the reflection coefficients, k_i of Figure 4.4, and predictor coefficients, a_i of Figure 4.5 (explained in the next section), and they represent the same information in the LP analysis [137, 105].

From either the direct-form filter realization or the mathematical derivation of lossless tube model [137, 105], linear prediction analysis is based on the all-pole filter:

$$H(z) = \frac{1}{A(z)} \quad \text{and} \quad A(z) = 1 - \sum_{k=1}^{p} a_k z^{-k} \qquad (4.9)$$

where $\{a_k, 1 \le k \le p\}$ are the predictor coefficients, and p is the order

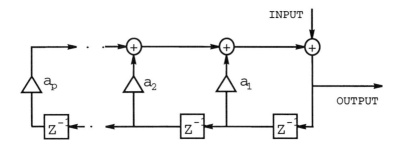

FIGURE 4.5
Direct form of all-pole filter representing vocal tract.

of the filter.

By transforming to the time domain, it can be seen that the system
of Equation 4.9 predicts a speech sample based on a sum of weighted
past samples:

$$s'(n) = \sum_{k=1}^{p} a_k s(n-k) \qquad (4.10)$$

where $s'(n)$ is the predicted value based on the previous values of the
speech signal $s(n)$.

4.2 Estimation of LP Parameters

To utilize the LP model for speech analysis, it is necessary to estimate
the LP parameters for a segment of speech. The idea is to find the a_ks
so that Equation 4.10 provides the closest approximation to the speech
samples, that is, so that $s'(n)$ is closest to $s(n)$ for all the values of n in
the segment. For this discussion, the spectral shape of $s(n)$ is assumed
to be stationary across the frame, or short segment of speech.

The error between a predicted value and the actual value is:

$$e(n) = s(n) - s'(n) \qquad (4.11)$$

substituting,

$$e(n) = s(n) - \sum_{k=1}^{p} a_k s(n-k) \qquad (4.12)$$

The values of a_k can be computed by minimizing the total squared error E over the segment:

$$E = \sum_n e^2(n) \qquad (4.13)$$

By setting the partial derivatives of E with respect to the a_ks to zero, a set of equations results that minimizes the error. Two solutions to the equations are presented below.

4.2.1 Autocorrelation Method of Parameter Estimation

For the autocorrelation method [105], the speech segment is assumed to be zero outside the predetermined boundaries. The range of summation of Equation 4.13 is $0 \leq n \leq N + p - 1$. The equations for the a_ks are compactly expressed in matrix form as:

$$\begin{bmatrix} r(0) & r(1) & \cdots & r(p-1) \\ r(1) & r(2) & \cdots & r(p-2) \\ \vdots & \vdots & \ddots & \vdots \\ r(p-1) & r(p-2) & \cdots & r(0) \end{bmatrix} \begin{bmatrix} a_1 \\ a_2 \\ \vdots \\ a_p \end{bmatrix} = \begin{bmatrix} r(1) \\ r(2) \\ \vdots \\ r(p) \end{bmatrix}$$

where $r(l)$ is the autocorrelation of lag l computed as:

$$r(l) = \sum_{m=0}^{N-1-l} s(m)s(m+l) \qquad (4.14)$$

and N is the length of the speech segment $s(n)$.

Because of the Toeplitz structure (symmetric, diagonals contain same element) of the matrix, the efficient Levinson-Durbin [109, 105] recursion can be used to solve the system. The equations are:

$$E^{(0)} = r(0) \qquad (4.15)$$

$$k_i = \frac{r(i) - \sum_{j=1}^{i-1} a_j^{(i-1)} r(i-j)}{E^{(i-1)}} \qquad (4.16)$$

$$a_i^{(i)} = k_i \qquad (4.17)$$

$$a_j^{(i)} = a_j^{(i-1)} - k_i a_{i-j}^{(i-1)}$$ (4.18)

$$E^{(i)} = (1 - k_i^2)E^{(i-1)}$$ (4.19)

where $1 \leq j \leq i - 1$. In all equations, i is the current order in the recursion, and the equations are solved in turn for all orders of $i = 1, 2, \cdots, p$.

The ith order coefficient of Equation 4.17 for values $1 \leq i \leq p$ is the ith reflection coefficient as discussed above. If the condition

$$|k_i| < 1 \qquad 1 \leq i \leq p$$ (4.20)

is met, the roots of the predictor polynomial will all lie within the unit circle in the z-plane, and the all-pole filter will be stable. Filter stability can be determined by checking this condition of the reflection coefficients.

4.2.2 Covariance Method

In the covariance method [3], the range of the summation of Equation 4.13 is limited to the range of the indices in the speech segment. This formulation results in the solution of the error minimization as:

$$\begin{bmatrix} c(1,1) & c(1,2) & \cdots & c(1,p) \\ c(2,1) & c(2,2) & \cdots & c(2,p) \\ \vdots & \vdots & \ddots & \vdots \\ c(p,1) & c(p,2) & \cdots & c(p,p) \end{bmatrix} \begin{bmatrix} a_1 \\ a_2 \\ \vdots \\ a_p \end{bmatrix} = \begin{bmatrix} c(1,0) \\ c(2,0) \\ \vdots \\ c(p,0) \end{bmatrix}$$

where the covariance c is:

$$c(i,k) = \sum_{m=0}^{N-1} s(m-i)s(m-k)$$ (4.21)

and includes values of $s(n)$ outside the original segment range of $0 \leq n \leq N - 1$.

Although the form for the covariance method is not Toeplitz, and does not allow the Levinson-Durbin recursion solution, efficient methods such as the Cholesky decomposition [105] can be used to solve the system of equations.

4.3 Transformations of LP Parameters for Quantization

Two transformations of the LP information have proven useful for coding. The log area ratios reduce the sensitivity to quantization noise when the value of the reflection coefficient is near 1. The line spectral frequencies (LSFs) are an ordered set of parameters, particularly suited to efficient vector quantization.

4.3.1 Log Area Ratios

The log area ratios are computed from the reflection coefficients as:

$$L_i = log\frac{1 + k_i}{1 - k_i} \tag{4.22}$$

and the inverse transform follows as:

$$k_i = \frac{1 + e^{L_i}}{1 - e^{L_i}} \tag{4.23}$$

4.3.2 Line Spectral Frequencies

In recent coder implementations, line spectrum pairs (LSPs), or line spectrum frequencies (LSFs), are the favored format for the LP parameter representation. The LSFs are the roots of the $P(z)$ and $Q(z)$ polynomials, where they are defined as:

$$P(z) = A(z) + z^{-(p+1)}A(z^{-1}) \tag{4.24}$$

$$Q(z) = A(z) - z^{-(p+1)}A(z^{-1}) \tag{4.25}$$

where $A(z)$ is the inverse LP filter of Equation 4.9, and p is the order of the LP analysis.

The p roots, or zeros, of $P(z)$ and $Q(z)$ lie on the unit circle, in complex conjugate pairs (in addition, one root will be at +1, and one at −1). Their angle in the z-plane represents a frequency, and pairs, or groups of three, of these frequencies are responsible for the formants in the LP spectrum. The bandwidth of the formant (how sharp the formant peak is) is determined by how close together the LSFs are for that formant.

Closer LSFs produce a sharper formant peak. This property provides a useful, practical check for stability after the LSFs have been quantized. The LSFs can be checked for a minimum spacing, and separated slightly if necessary.

Another desirable property of the LSFs is the localized nature of their spectral impact. If one LSF is adversely altered by the quantization and coding process, that will only degrade the LP spectrum near that LSF frequency. Other representations of the LP information (reflection coefficients, log area ratios) are not localized in frequency.

In practice, the zeros of the polynomials are found by numerical methods. Reference [86] provides a method to compute the LSFs using Chebyshev polynomials. Additional information on the properties of LSFs can be found in [151].

The LP coefficients, a_is, can be recovered from the LSFs by multiplying out the terms of the roots of Equations 4.24 and 4.25 (the LSFs) to obtain $P(z)$ and $Q(z)$. Then, $A(z)$ can be determined by noting that:

$$A(z) = \frac{1}{2}[P(z) + Q(z)] \qquad (4.26)$$

4.4 Examples of LP Modeling

For speech coding, the LP analysis models the shape of the short-term spectrum (frequency response of the vocal tract) for the purpose of efficient coding. The order of the LP analysis, p, is usually in the range of 8 to 14, with 10 being most common for coding applications. Higher model orders, 12 and above, accurately model the formant structure of voiced speech. But, the improved accuracy comes at the cost of more model parameters and the accompanying increase in bit rate necessary to encode the parameters.

Two example plots of the log of the magnitude of LP spectra along with the corresponding log magnitude DFT spectra are shown in Figures 4.6 (voiced) and 4.7 (unvoiced). In both cases, the order of the LP modeling was 12. Both the DFT and LP predictor coefficients were estimated from a 25 ms segment of speech for both figures. The LP predictor coefficients were computed using the autocorrelation method.

The LP spectrum is computed as:

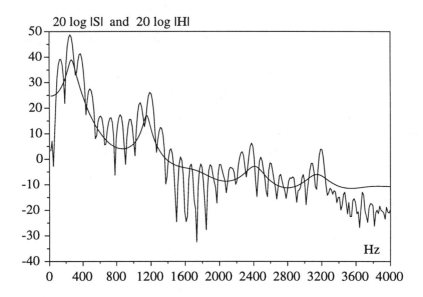

FIGURE 4.6
Log magnitude of DFT and LP spectra for a segment of voiced speech.

$$H_{LP}(\omega) = \frac{\sigma}{|A(\omega)|} \tag{4.27}$$

where σ is the square root of the energy of the segment, and $A(\omega)$ is defined in Equation 4.9. The $|A(\omega)|$ is computed as the magnitude of the DFT of the sequence $a(n) = 1 - a_1 - a_2 - \cdots - a_{p-1} - a_p$.

The plots indicate how the LP analysis models the general shape of the spectrum, but does not model the fine structure. In the voiced example, the LP representation does not model the pitch harmonics. The formants are evident in the LP spectrum of Figure 4.6 at approximately 300, 1200, 2400, and 3200 Hz. In Figure 4.7, the LP spectra models the overall vocal tract shape but does not model the random, noise-like fine structure displayed in the unvoiced DFT spectra. A prominent formant is evident at about 2900 Hz.

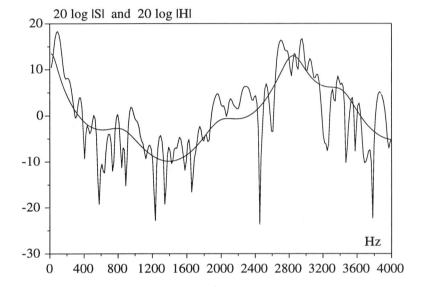

FIGURE 4.7
Log magnitude of DFT and LP spectra for a segment of un-voiced speech.

Chapter 5

Pitch Extraction

Pitch is a fundamental property of voiced speech. For voiced speech, the glottis opens and closes in a periodic fashion, imparting a periodic character to the excitation. The pitch period, T_0, is the time span between sequential openings of the glottis. The pitch frequency, F_0, is the reciprocal of the pitch period ($F_0 = \frac{1}{T_0}$). In this discussion, pitch and fundamental frequency are used interchangeably. Here, *pitch* does not mean the perceived, subjective tonal quality of a complex sound.

The range of fundamental frequencies for human speakers is 50 to 300 Hz. Men generally have pitch frequencies occupying the low portion of this range while women and children generally have pitch frequencies at the high end of the range. Fundamental frequencies are restricted to this range due to the physical limitations of the human vocal cords.

Much of the prosodic information in an utterance is carried by the rise and fall of the pitch. The ear is more sensitive to changes of fundamental frequency than to changes of other speech signal parameters by an order of magnitude [70]. As such, the quality of coded speech is highly influenced by an accurate regeneration of this parameter in the decoded output speech.

Estimating the pitch is more difficult than one might imagine. Pitch period estimates from the acoustic waveform can vary because the voiced excitation of the vocal tract is only quasi-periodic. Not only does the excitation waveform period vary slightly from one period to the next, but the time point chosen for period measurement will impact the pitch period – a peak-to-peak measurement will differ from a valley-to-valley measurement. The vibration of the vocal cords can even be quite nonperiodic, particularly at voicing onsets or the end of a phrase. Harmonics or subharmonics of the pitch frequency can appear more prominent than those of actual pitch frequency. These are just some of the variations that make pitch extraction an imprecise operation.

Literally hundreds of different pitch extraction methods and algorithms have been developed, but only a few will be discussed here. For a detailed survey of methods of determining pitch, please refer to [70]. Pitch extraction algorithms attempt to locate the periodicity in either the time-domain speech signal (or some preprocessed version of it) or a frequency-domain transformation of the speech. Autocorrelation methods and the variants cover the most popular approaches. Many speech coders begin estimating the pitch with an autocorrelation calculation. The initial estimate is often further refined to reduce occurrences of pitch halving and doubling errors.

5.1   Autocorrelation Pitch Estimation

The autocorrelation function is frequently used for pitch extraction. A correlation function is a measure of the degree of similarity between two signals. The autocorrelation measures how well the input signal matches with a time-shifted version of itself. The maxima of the autocorrelation function occur at intervals of the pitch period of the original signal.

The short-time autocorrelation function of a segment, $s(m)$, of a discrete-time signal, $s(n)$, is defined as:

$$r(k) = \sum_{m=0}^{N-1-k} s(m)s(m+k) \tag{5.1}$$

for the kth "lag," where N is the length of the segment. The signal $s(m)$ is assumed to be zero outside the range $0 \leq m \leq N - 1$. The change of index variables from n to m allows the segment to be indexed from 0 to $N - 1$, irrespective of the range of values of n for the segment.

Figure 5.1 displays a 200 sample segment of a voiced speech signal in the top waveform. (The sampling rate is 8000 Hz; the segment is 25ms long.) The autocorrelation of the segment is shown in the lower plot of Figure 5.1.

From the speech waveform, it can be seen that the pitch period is about 80 samples. Correspondingly, there is a prominent peak at lag 81 in the autocorrelation. A lag of 81 samples corresponds to a time period of 10.125 ms (81 samples/8000 Hz) and a pitch frequency of 98.8 Hz.

The maximum value for the autocorrelation will alway be at lag 0 (no shift). This equates to $k = 0$ in Equation 5.1, and is the computation

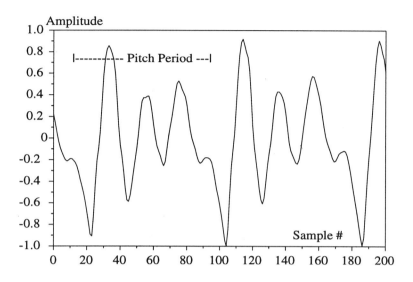

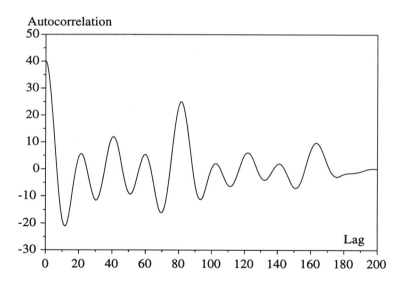

FIGURE 5.1
Time-domain waveform and autocorrelation of a short segment of voiced speech.

for the energy of the speech segment. A smaller local maximum appears at lag 162, which indicates the good match when the shift is twice the pitch period. For pitch extraction, the window of speech should contain at least two pitch periods ($N > 2/F_0$) to allow the first to match up with the second at the shift equal to the pitch period.

5.1.1   Autocorrelation of Center-Clipped Speech

Because speech is not a purely periodic signal and vocal tract resonances produce additional maxima in the autocorrelation, pitch analysis on a direct autocorrelation of the speech signal can result in multiple local maxima. The maxima corresponding to the true pitch period can be difficult to discern. There are several methods to suppress these local maxima (which can usually be attributed to the damped oscillations of the vocal tract response to a voiced excitation). Sondhi [150] suggested the method of center clipping the speech before computing the autocorrelation. The center-clipped speech is obtained by the nonlinear transformation:

$$y(n) = C[s(n)] \tag{5.2}$$

$C[]$ is shown in Figure 5.2 and C_L is set as a fixed percentage of the maximum amplitude of the speech signal (Sondhi [150] used 30%).

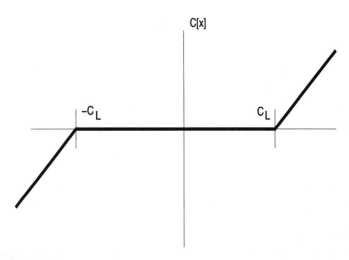

FIGURE 5.2
Center clipping function.

For samples with amplitude above C_L, the output of the center clipper is equal to the input minus the clipping level. For samples with magnitude below the clipping level, the output is zero.

Figure 5.3 shows a diagram of a center-clipped speech segment (the segment of Figure 5.1) and the autocorrelation function of the clipped waveform. The autocorrelation shows that the peak corresponding to pitch period is prominent, while the other local maxima have been reduced. The peak of the autocorrelation of the center-clipped speech is much more distinguishable than in the autocorrelation of the original speech.

While the center-clipping operation enhanced the performance of the autocorrelation in this example, center clipping can reduce the effectiveness under less ideal cases. If the signal is noisy or only mildly periodic (voice onsets), the clipping operation might remove beneficial signal information. For segments of rapidly changing energy, setting an appropriate clipping level can be difficult, even if it is adjusted dynamically.

5.1.2 Cross Correlation

The autocorrelation calculation of Equation 5.1 includes fewer terms as the lag increases because of the subtraction of k from the upper limit of the summation. This effect can be seen in the roll off of the high lag values in the lower plot of Figure 5.1. These high lag values of the autocorrelation are important for low-pitched male speakers. For a 50 Hz pitch, the lag (number of samples) between successive pitch pulses is 160 samples at an 8000 Hz sampling rate. The cross correlation offers an alternative computation without this limitation.

The cross correlation operates on two separate data windows, each of length N. Each value of the correlation is summed over the same number of terms, N. The cross correlation is computed as:

$$c(k) = \sum_{m=0}^{N-1} s(m)s(m+k) \qquad (5.3)$$

The only difference from Equation 5.1 is the upper limit on the summation. However, this is best interpreted as two separate data segment windows, one for each factor of s in the equation. This is most distinctive for the case of $k > N$, where each segment in the summation, originating from $s(n)$, will cover separate, nonoverlapping ranges of n.

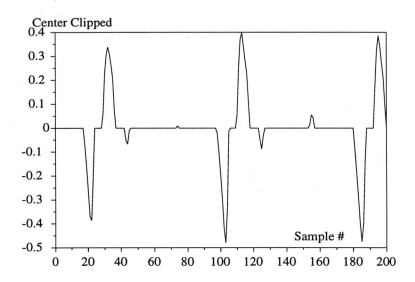

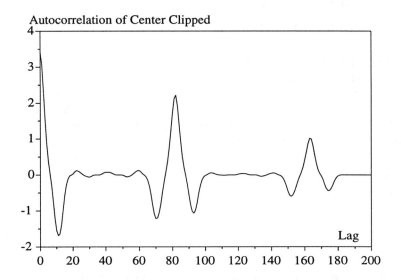

FIGURE 5.3
**Center-clipped waveform and autocorrelation of a short seg-
ment of voiced speech.**

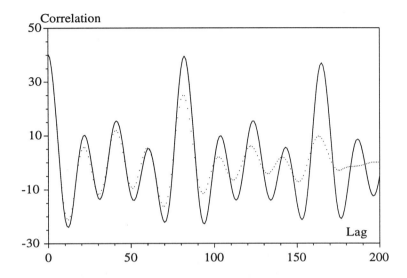

FIGURE 5.4
Cross correlation (solid) and autocorrelation (dotted) of the voiced speech segment of Figure 5.1.

Here, the terminology *cross correlation* is used to reflect the two distinct signal segments. In some texts [125], the computation of Equation 5.3 is referred to as the *autocovariance* because both segments originate from the same signal. More precisely, for computation of the autocovariance, the signal segments first have their respective mean values subtracted (yielding zero mean segments) before the multiplication and summation of Equation 5.3.

Figure 5.4 displays the cross correlation and autocorrelation for the 200 sample speech segment of Figure 5.1. The roll off of the autocorrelation is apparent for higher lag values. The cross correlation does not diminish at high values, and the local maxima near lag 163 corresponds to twice the pitch period.

For a segment length of $N = 200$ and to compute lags out to $k = 200$ as shown in the figure, 400 speech samples will be required. Therefore, the computation will include 200 more samples than are displayed in the plot. This is not required for adequate performance; window lengths of near one pitch period will perform well, and the window size is not tied to the range of lag values computed.

5.1.3   Energy Normalized Correlation

Figure 5.4 illustrates a potential problem with the cross correlation. Because the correlation values at $1/F_0$ and $2/F_0$ are nearly the same, the correlation value at $2/F_0$ could be larger than that at $1/F_0$ during segments of increasing energy. Rapidly increasing energy is common at voicing onsets. If the correlation values are compensated based on the energy in the sliding window (shifted by k), the correlation will match the shape, but not vary depending on the energy.

The normalized correlation is expressed as:

$$c_{norm}(k) = \frac{\sum_{m=0}^{N-1} s(m)s(m+k)}{\sqrt{\sum_{m=0}^{N-1} s^2(m)}\sqrt{\sum_{m=0}^{N-1} s^2(m+k)}} \qquad (5.4)$$

where the energy terms have been added to the denominator.

The top plot of Figure 5.5 displays a speech segment of increasing energy. The bottom plot shows the cross correlation and cross correlation normalized by the energy as in Equation 5.4. Both plots have been normalized to place their maximum values at 1.

The largest local maximum for the unnormalized cross correlation occurs at twice the pitch period, at a lag of about 170, due to the increasing energy. The normalized cross correlation displays a higher local maxima at lag 85, the true pitch period, than at lag 170.

The normalized cross correlation estimate for the pitch is among the most popular methods of pitch estimation.

5.2   Cepstral Pitch Extraction

When a periodic signal with fundamental frequency F_0 consists of many adjacent harmonics (as voiced speech signals do), the corresponding short-term spectrum exhibits a ripple due to its harmonic structure. This is particularly evident as Noll showed for the logarithmic power spectrum where this ripple takes on a cosine-like shape [122]. The ripple of the harmonic structure is evident in any of the plots of voiced speech spectra in Chapter 2. The *cepstrum* of this signal will exhibit a strong peak at *quefrency d* (defined below) equal to the period duration $1/F_0$.

The cepstrum is defined as the inverse discrete Fourier transform of the log of the magnitude of the discrete Fourier transform of the input

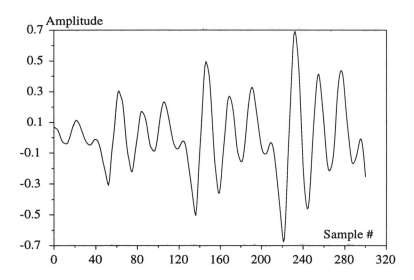

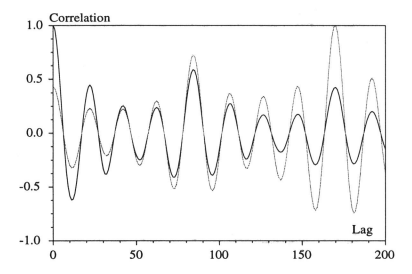

FIGURE 5.5
Increasing energy speech segment in top plot, and cross correlation (gray) in bottom plot and normalized cross correlation
(black).

signal $s(n)$. The inverse DFT and DFT are defined in Equations 3.21 and 3.20. In symbolic notation, the cepstrum is expressed as:

$$\text{Cepstrum}(d) = \text{IFFT}(\log_{10}|\text{FFT}(s(n))|) \qquad (5.5)$$

The index d is defined as the quefrency of the cepstrum signal. Because of the transform and inverse transform, quefrency is a type of time-domain index. A peak in the cepstrum at quefrency d_o corresponds to a periodic component in the original signal with period d_o and frequency $1/d_o$.

The cepstrum extracts pitch information from a voiced speech signal because a voiced signal not only contains dominant spectral components at the fundamental frequency, but also contains harmonics of the pitch fundamental. The cepstrum captures the repeated structure in the magnitude of the spectrum. The low quefrency range of the cepstrum represents general vocal tract shape. The higher quefrency portion of the cepstrum represents the excitation information and, in the case of voiced speech, the pitch.

Figure 5.6 displays the log magnitude spectrum and the corresponding cepstrum for the speech segment of Figure 5.1. The large value at Cepstrum(0) (the "DC" value) has been removed to better resolve the dynamic range of the plot. The prominent peak at about quefrency 82 indicates the pitch periodicity. Here, the quefrency is in samples and corresponds to the sampling rate of the original signal, 8000 Hz. Therefore, a quefrency of 82 translates to a pitch frequency of $8000/82 = 97.6$ Hz.

The significant structure in the low quefrency range, from 1 to about 16, represents the vocal tract information. In fact, the low quefrency cepstral values have been suggested as a compact vocal tract representation in [125].

A Cepstral analysis of a short time segment of speech will produce a peak at the pitch period for voiced speech, but no prominent peaks for unvoiced speech. Cepstral analysis can be used to determine if a speech segment is voiced or unvoiced [125] and to determine the pitch period, $1/F_0$, if the segment is voiced.

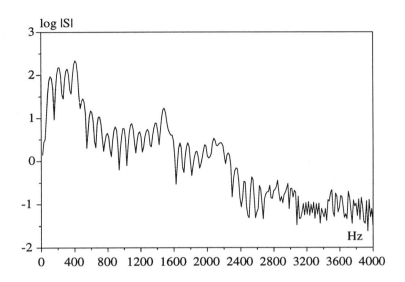

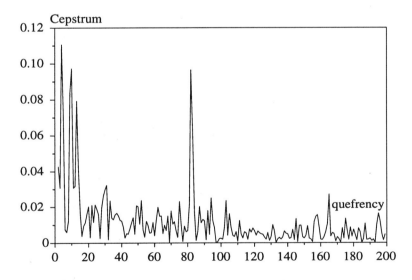

FIGURE 5.6
Log magnitude of DFT and cepstrum of speech segment of
Figure 5.1.

5.3   Frequency-Domain Error Minimization

Frequency-domain error minimization [65] is an analysis by synthesis approach to pitch estimation. An initial pitch estimate is used to construct a synthesized speech spectrum based on a harmonic speech model. The synthesized spectrum is compared to the original, and the pitch estimate is adjusted based on the error.

Frequency-domain error minimization is an iterative method that yields accurate pitch period estimates at the cost of computation complexity. Due to the high computational cost of this method, it is often used in conjunction with computationally simple methods such as the autocorrelation method. The simpler method narrows the pitch period range to its closest integer sample, and then frequency-domain error minimization is used to "fine tune" the estimate to the desired accuracy.

The algorithm adjusts the pitch estimate to minimize the error in the reconstructed speech. The error between the original spectrum and the synthesized spectrum is:

$$\epsilon = \int_{-\pi}^{\pi} |S_w(\omega) - \hat{S}_w(\omega)|^2 d\omega \tag{5.6}$$

where $S_w(\omega)$ is the Fourier transform of the original windowed speech segment, and $\hat{S}_w(\omega)$ is the Fourier transform of the synthesized speech.

The synthesized spectrum can be expressed as:

$$\hat{S}_w(\omega) = \sum_{k=1}^{K} A_k W(\omega - k\omega_o) \tag{5.7}$$

where A_k is the amplitude at the kth harmonic, ω_o is the pitch estimate ($\omega_o = 2\pi F_o$), and $W(\omega)$ is the Fourier transform of the time domain window. This expression can be visualized as the frequency response of the window, shifted in frequency to each harmonic, and scaled by the magnitude of that harmonic.

For a given pitch period, $T_o = 2\pi/\omega_o$, the best spectral magnitudes, yielding smallest error, ϵ, at harmonics of the pitch frequency, are calculated by: [65]

$$A_k(\omega_o) = \frac{\int_{k-\frac{1}{2}}^{k+\frac{1}{2}} S(\omega)W(\omega - k\omega_o)d\omega}{\int_{k-\frac{1}{2}}^{k+\frac{1}{2}} W^2(\omega - k\omega_o)d\omega} \tag{5.8}$$

This expression integrates the energy in the original spectrum over the frequency range of one harmonic, and normalizes by the window shape over that frequency range.

Given ω_o and the magnitudes, A_k, the synthesized spectrum can be calculated. A search procedure is employed to refine the pitch estimate by minimizing the error given by Equation 5.6.

5.4 Pitch Tracking

Most of the time, the pitch varies smoothly over time. Sometimes it changes significantly and abruptly. When the change is smooth and slow, good-sounding synthesized speech depends on accurate representation of this smoothness. However, abrupt jumps of the pitch frequency are equally important to maintaining natural sounding synthesized speech. Pitch period doubling, caused by the type of ambiguity displayed in the correlation plots of Section 5.1.3, can introduce large abrupt pitch changes that are incorrect. These conflicting requirements necessitate post processing of the time evolution of the pitch estimates. The time evolution of pitch values is called the *pitch track*. Two methods have been employed most frequently: median smoothing [136] and dynamic programming [120].

5.4.1 Median Smoothing

Median smoothing can be considered as a filter of length L. The input to the filter is L sequential values of the pitch track. The output of the filter is simply the median of the L values. The median is the "middle" value, with $(L-1)/2$ values greater and $(L-1)/2$ values less than the median. The length of L is usually 3 or 5. Median filtering can effectively remove single or double occurrences of pitch halving and doubling. A linear lowpass filter can be used after median filtering to make the pitch track smoother by reducing small-level spurious noise in the estimates. Median smoothing is also discussed in [137].

5.4.2 Dynamic Programming Tracking

Dynamic programming (DP) [9] encompasses a large class of optimization algorithms applicable to many different fields. In a general sense, DP can be thought of as optimizing a sequence of decisions through a network, each based on the local constraint at the current node and the current state. This method offers significantly reduced computation over selecting the best solution after complete enumeration of the all possible solutions. Exhaustively checking all solutions is infeasible for many optimization problems.

As DP applies to pitch tracking [120], it entails optimizing the pitch track based on a sequence of decisions that selects from candidate estimates for the current frame based on tracks backward in time and a local constraint. DP is a method for finding the best (lowest *cost*) path through the pitch estimates. The candidate estimates are the pitch period values and their associated errors (correlation values at that pitch period in the case of a correlation estimator). Candidates can be either local maxima from the correlation or the entire correlation (all values at each computed lag). The *path cost* accumulates the error across a number of frames by adding in the error for the selected pitch candidate at each frame.

The local constraint is biased toward tracks that do not change rapidly, and toward lower pitch period values to reduce the chance of pitch period doubling. However, when the pitch does indeed jump significantly, this new track will be selected because as the track evolves, it will be seen to have an overall lower error, or path cost, associated with it.

For a given stage in the process, the best track is saved for *each* candidate of the current stage. For pitch tracking, these numerous "best tracks" typically converge to a single global best track in a relatively short time.

Chapter 6

Auditory Information Processing

A grasp of both the theory of speech production and the theory of human audition is essential to understand the fundamentals of speech coding. The fact that speech is generated through a human vocal tract allows a more compact signal representation for analysis/synthesis as opposed to a generic acoustic signal. Because decoded speech is synthesized for the human ear, further reductions in the signal representation are possible by disregarding signal information that cannot be perceived. Various components of the signal interact and interfere to determine the "perceived sound." These facts have been applied to high-fidelity coding of audio [82, 83, 84, 85] for the consumer electronic market, for Internet audio compression, and within standards such as MPEG-2 and MPEG-4 (see Appendix A). This processing can be incorporated into speech coders to further reduce the information needed to regenerate high-quality speech.

This chapter begins with a description of how the ear performs frequency analysis and continues with the concept of *critical bands*. The minimum detectable sound level is presented for quiet and noisy acoustic environments. This leads into *masking*, in both frequency and time. The information in this chapter provides the basis for perceptual speech coding. Chapter 12 describes how masking can be used to improve speech coding efficiency.

6.1 The Basilar Membrane: A Spectrum Analyzer

The basilar membrane is a key component of the inner ear. Oversimplified, sound vibrations cause movement of the basilar membrane by transduction through the middle ear. Movement of the basilar membrane stimulates hair cells, which in turn produce impulses in the auditory nerve fibers.

Ohm and Von Helmholtz [166] were the first to present the notion that the basilar membrane acts as a spectrum analyzer. Von Békésy expounded upon this theory and demonstrated that the basilar membrane vibrates locally, and the point of vibration is related monotonically to the frequency of the acoustic stimulus [164, 165]. Von Békésy proved that the basilar membrane was a spectrum analyzer, not an array of tuned resonators, but a nonuniform (almost logarithmically scaled) transmission line with limited but distinct spectral resolution. Further experimentation showed that this limited spectral resolution was characterized by *critical bands* [10].

6.2   Critical Bands

In loose terms, a critical band can be thought of as a frequency span, or frequency "bin," into which sounds are lumped perceptually. Although critical bands can be defined experimentally, the following definitions are most useful for the purposes of this discussion:

"The threshold [of audibility] [see Section 6.3] of a narrow band of noise lying between two masking tones remains constant as the frequency separation between the tones increases until the critical band is reached; then the threshold of audibility of the noise drops precipitously." [141]

"The loudness of a band of noise at a constant sound pressure remains constant as the bandwidth increases up to the critical band; then the loudness begins to increase." [141]

It is adequate to say that the critical band is a frequency range, defined by its band edges (specific frequencies), outside of which subjective responses change abruptly.

The fact that a critical band can be saturated is important to speech coding. In this case, *saturated* refers to the critical band being "filled" with sound, in that additional lower level sounds added to that frequency range cannot be perceived. The fact that certain acoustic stimuli cannot be sensed by the human ear is crucial because these stimuli need not be preserved for accurate coding. This savings allows coding resources to be allocated to frequency ranges where the sound will be perceived.

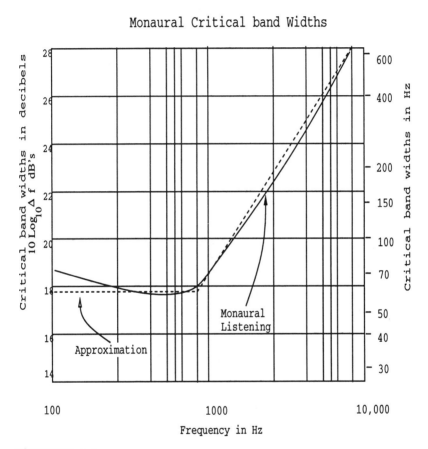

FIGURE 6.1
Frequency width of critical bands as a function of the band center frequency.

Critical Band No. (Barks)	Frequency (Hz)	Mels
1	20-100	0-150
2	100-200	150-300
3	200-300	300-400
4	300-400	400-500
5	400-510	500-600
6	510-630	600-700
7	630-770	700-800
8	770-920	800-950
9	920-1080	950-1050
10	1080-1270	1050-1150
11	1270-1480	1150-1300
12	1480-1720	1300-1400
13	1720-2000	1400-1550
14	2000-2320	1550-1700
15	2320-2700	1700-1850
16	2700-3150	1850-2000
17	3150-3700	2000-2150
18	3700-4400	2150-2300
19	4400-5300	2300-2500
20	5300-6400	2500-2700
21	6400-7200	2700-2850
22	7200-9500	2850-3050

Table 6.1 The relationship between the frequency units: Barks, Hertz, and Mels.

Extensive experimental research has been performed to quantify critical bandwidth as a function of the frequency at the center of the band. Figure 6.1 [10] shows the results of these experiments for single ear listening. As can be seen from this figure, at center frequencies greater than 500 Hz, critical bandwidth increases approximately linearly as center frequency increases logarithmically.

Figure 6.1 is the basis for the Bark domain and the Mel domain. Both the Bark and the Mel domains were created to have a constant number of each unit (Barks or Mels) in each critical band. The Bark domain was normalized to have 1 Bark per critical band. Barks and Mels are perceptually based frequency units that increase, almost logarithmically, with frequency.

Table 6.1 illustrates the relationship between the frequency units of Barks, Mels, and Hertz. The table shows that each critical band contains

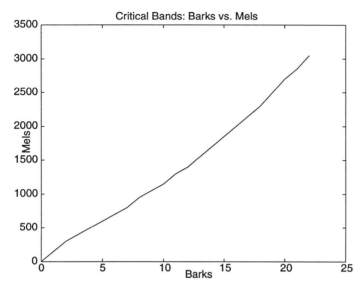

FIGURE 6.2
Comparing the experimentally derived frequency scales of
Barks versus Mels.

a logarithmically increasing frequency bandwidth in the linear scale of
Hertz. Approximately 150-200 Mels span each critical band. By definition, there is 1 Bark per critical band.

Figure 6.2 shows a graph of Barks versus Mels. Although the line
is somewhat linear, it is not exactly linear. This is because all of the
information known about critical bands is a result of experimental tests,
which are far from exact. The fact that both units are so close to being
linearly related even though they are formed on the basis of separate
experimental tests, supports the validity of these frequency scalings.

6.3 Thresholds of Audibility and Detectability

The *threshold of audibility* for a specified acoustic signal is the minimum effective sound pressure that is capable of evoking an auditory
sensation in the absence of noise in a specified fraction of the trials [10].
It is often expressed in decibels relative to 0.0002 microbar, which is

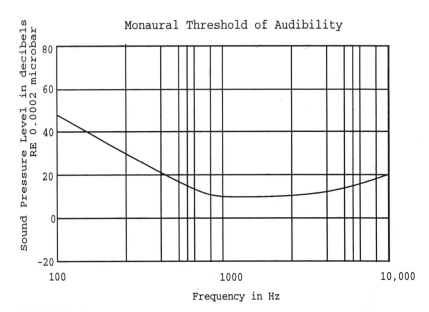

FIGURE 6.3
Threshold of audibility for a pure tone in silence.

considered the absolute threshold of audibility in terms of pressure.

The American standard threshold of audibility for monaural hearing of pure tones, for a listener with normal hearing seated in an anechoic (echo-free) chamber wearing earphones, is shown on the curve of Figure 6.3. The sound pressure is measured at the entrance to the ear canal. In other words, a person with "normal" hearing cannot hear tones below the curve (softer) but can hear tones above the curve (louder). The term *normal hearing* is used because some people have better than normal hearing (and can hear some tones below the curve) and some have subnormal hearing (and conversely cannot hear some tones above the curve).

The *threshold of detectability* for a specified acoustic signal is the minimum effective sound pressure that is capable of evoking an auditory sensation in a specific acoustic environment. Therefore, the *threshold of detectability* is identical to the *threshold of audibility* when the specific acoustic environment is silence, and conversely, the *threshold of detectability* is highly elevated in the specific acoustic environment of a crowded, noisy restaurant.

6.4 Monaural Masking

It is much more difficult to hear a specific sound in noisy surroundings than to hear that same sound in a quiet environment. One needs to shout to make oneself heard in a crowded restaurant, but in the silence of a library, a gentle whisper can often disturb others. Psychophysicists have learned a great deal about how the ear analyzes sounds by examining the way certain sounds drown out, or *mask*, other sounds [29].

One of the most valuable, and exploitable, properties of hearing is that of *monaural masking* [46]. *Masking* is defined as "the process by which the detectability of one sound (the *maskee*) is impaired by the presence of another sound (the *masker*)." [28]

6.4.1 Simultaneous Masking in Frequency

Simultaneous masking is masking where both sounds (*the maskee and the masker*) occur at the same instance in time. In-depth studies have been done on simultaneous masking of pure tones on a pure tone [46, 43, 81, 139, 45, 140, 177, 89, 71]. If a tone is sounded in the presence of a strong tone close in frequency (particularly if it is in the same critical band, but this is not essential), its *threshold of detectability* is substantially elevated as shown in Figure 6.4 [81].

The figure shows the tones and levels that a normal listener can hear in the presence of a 1200 Hz, 80 dB primary tone. All weaker signals below the curve cannot be heard by a normal listener. Notice that the masking effect is much more prevalent when the secondary tone is at a frequency greater than the primary tone. Also note that the masking effect is strongest when the secondary tone is very close in frequency to the primary tone.

Similar results are observed when one or both of the sounds are bands of noise [155]. Therefore, in a complex spectrum of sound, some weak components in the presence of stronger ones are not detectable at all. Spectral analysis and examination of *simultaneous masking in frequency*, carried out moment by moment, forms the basis of current algorithms for efficient coding of wideband audio and can be utilized for efficient coding of speech.

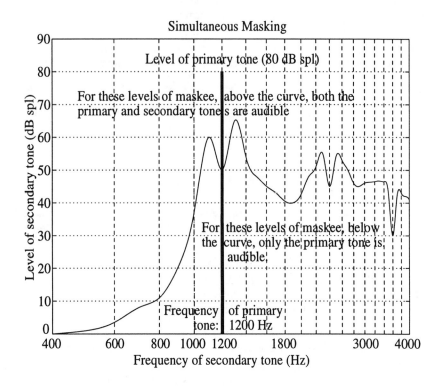

FIGURE 6.4
**Simultaneous masking in frequency of one tone on another tone
(data adapted from [81]).**

6.4.2 Temporal Masking

Masking can occur between signals that are separated in time and nonoverlapping. A loud sound followed closely in time by a weaker one can elevate the threshold of detectability of the weaker one and render it undetectable. Surprisingly, the masking effect works when the weaker sound is presented prior to the stronger sound, but too a much lesser extent. The fact that the masker can occur later or earlier than the maskee gives rise to the terminology *forward and backward temporal masking* [34]. A great deal of experimentation has also been done to characterize the temporal qualities of masking [176, 72, 31, 28, 134, 34, 81, 155, 140, 89, 141, 139].

Figure 6.5 illustrates both forward and backward masking. If a primary signal occurs at time t_0 and a secondary signal of the same frequency occurs at time $t_0 + \Delta$, then the secondary signal cannot be heard if the amplitude difference of the two tones is less than the threshold indicated in the curve. For example, if a 1200 Hz, 80 dB sound pressure level (spl) primary tone is present at time $t_0 = 0$ and a 1200 Hz, 30 dB spl secondary tone is present at time t = 30ms, then the secondary tone is completely masked because 30 dB < (80 dB - 38 dB). Similar calculations can be performed using the curve for backward masking, but the secondary tone occurs at a time Δ *before* the primary tone.

For maximum coding bit savings, both simultaneous frequency and temporal masking are considered together. Chapter 12 describes the application of masking to perceptual coding.

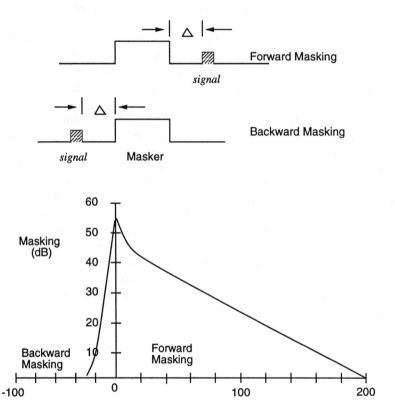

△ - Time Between Signal and Masker (msec)

FIGURE 6.5
Illustration of the effect of temporal masking.

Chapter 7

Quantization and Waveform Coders

The goal of quantization is to accurately encode data using as little information (as few bits) as possible. Efficient and accurate parameter quantization is central to speech coding because pertinent information must be represented as accurately as the coding requirements dictate using as little information as possible. Quantization can be applied directly to a sampled speech waveform or to parameter files such as the output of a vocoder analysis.

Waveform coders encode the shape of the time-domain waveform. Basic waveform coding approaches often do not exploit the constraints imposed by the human vocal tract on the speech waveform. As such, waveform coders represent nonspeech sounds (music, background noise) accurately, but do so at a higher bit rate than that achieved by efficient speech-specific vocoders.

Vector quantization (VQ) encodes groups of data simultaneously instead of individual data values. Advances in vector quantization of line spectral frequencies (LSFs) is one of the primary reasons for improved speech quality in leading low bit-rate coding schemes.

This chapter covers basic quantization of single element data and various waveform coding approaches. VQ is presented along with the computation reduction techniques that make it practical. The chapter concludes with a description of current approaches for efficient quantization of LSFs.

7.1   Uniform Quantization

The simplest type of quantization is uniform, or linear, quantization. The range of values for the signal is segmented into evenly spaced quantization levels. The number of levels is equal to the number of codewords available for quantization. If the capacity of n bits is used, there are 2^n codewords available and 2^n quantization levels. A codeword directly represents a quantized level of the signal.

7.1.1   Uniform Pulse Code Modulation (PCM)

When *uniform quantization* is applied directly to an audio waveform, the process is called *pulse code modulation* (PCM). Pulse code modulation is the simplest method of speech coding and is essentially the sampling process as discussed in Section 3.1. An analog speech signal is sent into an anti-aliasing analog lowpass filter which eliminates all frequencies above half the sampling rate. The signal is then sent through an analog-to-digital (A/D) converter which converts the signal to a sequence of numbers, with the time distance between sample points equal to the sampling rate. The signal, now a sequence of numbers, can be stored or sent through a digital transmission channel.

The PCM analysis process is displayed in Figure 7.1. The input signal is an analog signal, typically a varying voltage level in an analog circuit. The lower plot of the input signal represents the continuous-time frequency domain of the input speech. The second plots (time domain upper, frequency domain lower) display the continuous-time impulse and frequency responses, respectively, of the analog low-pass filter. The input speech is bandlimited by the lowpass filter with the result displayed in the third plots. The bandlimited analog signal is sampled at discrete time intervals to produce the last plots. The samples are shown as the dots on the time domain waveform. The frequency domain plot indicates the cyclical nature of the Fourier representation of a discretely sampled signal.

To reconstruct the analog signal, the digital signal is passed through a Digital-to-Analog (D/A) converter and then filtered by a simple low-pass interpolating analog filter which generally has the same characteristics as the anti-aliasing pre-filter that was used to filter the original analog signal. A representation of the PCM reconstruction process can be seen in Figure 7.2. The discretely sampled signal is converted to the

Time Domain Representation

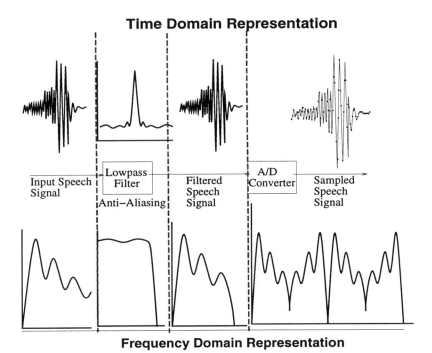

Frequency Domain Representation

FIGURE 7.1
Time- and frequency-domain representations of signals at different stages during pulse code modulation (PCM) analysis.

pulse-type waveform of the second plots. This waveform has higher harmonics not present in the original signal. The lowpass filter removes these unwanted higher frequencies.

PCM is a simple coding scheme and is often used when transmission bandwidth (or storage space) is not a limitation. PCM is more susceptible to bit errors than other speech waveform coding methods such as delta-modulation [79], because a single bit error can change a value from the positive maximum value to the minimum value possible. Therefore, if speech quality is important in a noisy transmission environment, PCM is not desirable even if the coding bit rate is not an issue.

The *reconstruction error* (the difference between the original signal and the reconstructed signal) is affected by quantization error that is introduced in the PCM coding scheme. This error is introduced during the process of analog-to-digital conversion. In order to represent a signal digitally, the values of the signal must be approximated to the

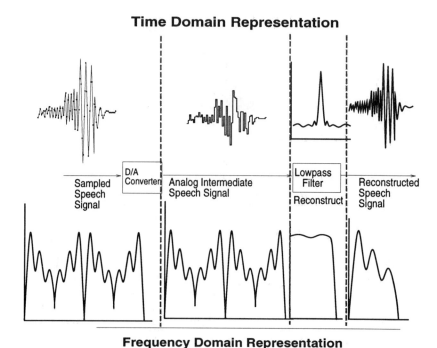

FIGURE 7.2
Time- and frequency-domain representations of signals during pulse code modulation (PCM) reconstruction.

closest possible discrete values. For example, if the A/D converter represents each value with only 3 bits, then the *dynamic range* of the signal is sectioned into 8 even parts and each sample is represented by the closest match. For a signal that fluctuates through the range of [-1V to 1V], its (*dynamic range*) is represented by the values: [-7/8 -5/8 -3/8 -1/8 1/8 3/8 5/8 7/8]. The quantization error introduced by a 16 bit D/A converter is generally not perceptible to the human ear. For a speech signal that may contain some pure silence, one may choose a quantization scheme that contains zero as a quantization value, so that quantization noise is not introduced when no signal is present. This can be accomplished by shifting the values by $1/2^B$ in either direction where B is the number of bits per sample used in quantization. In the above example, the quantization values shift to [-1 -3/4 -1/2 -1/4 0 1/4 1/2 3/4]. The only issue with this scheme is that it is not symmetric: -1 V is represented, but 1V must be approximated by 3/4 V.

7.2 Nonlinear Quantization

It is often beneficial to use nonlinear spacing between the quantization levels. The spacing of the quantization levels can be set based on the distribution of sample values in the signal to be quantized. The distance between adjacent quantization levels is set smaller for regions that have a larger share of the sample values. When adjusted in this manner, the overall quantization error is smaller. In direct speech waveform coding, logarithmically spaced quantization levels are used to best match the expected distribution of the speech signal.

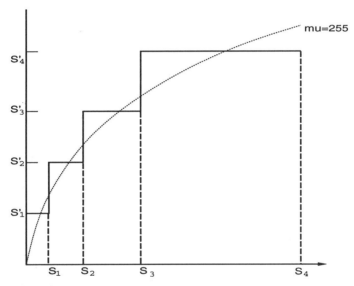

FIGURE 7.3
Distribution of quantization levels for a nonlinear 3-bit quantizer.

Figure 7.3 shows the distribution of quantization levels for a nonlinear 3-bit quantizer. The input value S is positioned on the x-axis, and the corresponding output value is S' on the y-axis. Although only four levels are shown, the same logarithmic scale is used for negative S, where the third bit indicates the sign of S and S'.

7.2.1   Nonuniform Pulse Code Modulation

Nonuniform pulse code modulation works similarly to PCM except that the quantization values are nonlinearly distributed through the dynamic range. Schemes employ fine quantizing steps around frequently occurring values, and course step sizes around the more rarely occurring values. An alternative view is to mimic the human ear and distribute the bits so that the quantization noise is less perceptible. μ-law and A-law coders fall under the latter of these methods. They are both quasi-logarithmic in that they are linear for small signal values and logarithmic for large signal values.

The formula for A-law companding (compressing/expanding) is:

$$c(x) = \begin{cases} \frac{A|x|}{1+log_e A}\text{sgn}(x); & 0 \leq \frac{|x|}{x_{max}} \leq \frac{1}{A} \\ x_{max}\frac{1+log_e(A|x|/x_{max})}{1+log_e A}\text{sgn}(x); & \frac{1}{A} < \frac{|x|}{x_{max}} \leq 1 \end{cases} \qquad (7.1)$$

where the signal x has the dynamic range: $-x_{max} \leq x \leq x_{max}$, $A \geq 1$, and sgn(x) represents the sign of x.

The formula for μ-law companding is:

$$c(x) = x_{max}\frac{log_e(1+\mu|x|/x_{max})}{log_e(1+\mu)}(sgn)(x); \qquad \mu \geq 0 \qquad (7.2)$$

A-law companding is used in European telephony (the European PCM standard) with A=87.56. North American telephony employs μ-law companding (The North American PCM standard) with μ=255. The A-law and μ-law standards are specified in the International Telecommunications Union (ITU) recommendation G.711 [186].

Figure 7.4 plots companding functions for A-law and μ-law for different values of A and μ. The bottom plot highlights the difference between the North American and European standards. As can be seen, the two are essentially the same, differing only slightly for very small input values.

7.3   Differential Waveform Coding

A coder that quantizes the difference waveform rather than the original waveform is called differential pulse code modulation (DPCM). Ei-

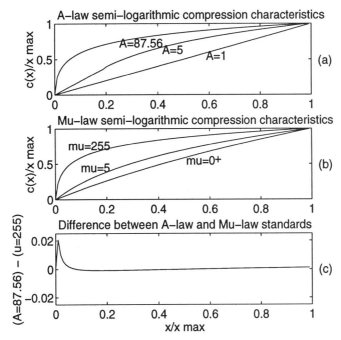

FIGURE 7.4
Companding functions for A-law and μ-law for different values
of A and μ. The bottom plot indicates the difference between
North American and European standards.

ther linear or nonlinear quantizers can be used in DPCM systems. First-
order or higher-order predictors can also be used to enhance DPCM
performance. The linear delta modulation quantization scheme is also a
type of DPCM coder.

As mentioned above, PCM, both uniform and nonuniform, is quite
susceptible to bit errors. If one of the more significant bits is erro-
neously reversed, the representation of that sample will be drastically
off. Differential waveform coders such as differential pulse code modulation (DPCM) and delta modulation produce less perceptual error for
single bit errors. These coders encode the difference signal (the difference
between adjacent samples) rather than the original signal. These methods yield poor performance if the signal is completely random; however,
because subsequent samples in speech signals are highly correlated, the
difference signal generally has a smaller dynamic range than the origi-

nal signal. As such, quantizing the difference signal yields better coding quality than uniform PCM at the same bit rate.

7.3.1   Predictive Differential Coding

Predictors are often used in differential quantization to lower the variance of the difference signal. The smaller the variance of the coded signal, the better the quality of the coding that can be achieved with all other variables being equal. Predictive differential coding predicts the value of the present sample from the values of previous samples, and then encodes the difference between the predicted and actual sample values. The input to a quantizer of this type is the difference signal:

$$d(n) = x(n) - \tilde{x}(n) \qquad (7.3)$$

which is the difference between the unquantized input sample, $x(n)$, and a predicted value of the input sample, $\tilde{x}(n)$.

If the prediction is accurate, then $\tilde{x}(n) \approx x(n)$ and the variance of the difference signal $d(n)$ is smaller than the variance of the original signal $x(n)$. A differential quantizer with a given number of levels yields a smaller quantization error than does quantization of the same highly correlated signal directly.

The predicted value $\tilde{x}(n)$ is often calculated using linear prediction (LP). That is, $\tilde{x}(n)$ is a linear combination of the past p quantized values:

$$\tilde{x}(n) = \sum_{k=1}^{p} a_k \hat{x}(n-k) \qquad (7.4)$$

The optimal values (minimum prediction error) for lowpass filtered speech for up to fifth-order prediction are as follows:

$$\begin{bmatrix} a_1 = 0.86 \\ a_2 = 0.64 \\ a_3 = 0.40 \\ a_4 = 0.26 \\ a_5 = 0.20 \end{bmatrix} \qquad \text{Values from[123]}$$

These prediction coefficients are determined by performing LP analysis on long time durations (many minutes) of the speech signal that include a representative distribution of different speech sounds. See Chapter 4 for a description of LP analysis.

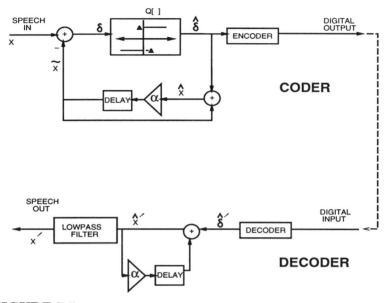

FIGURE 7.5
Delta modulation with 1-bit quantization and first-order pre-diction.

7.3.2 Delta Modulation

The simplest form of differential quantization is first-order, one-bit linear delta modulation. It has a single predictor ($a_1 = \alpha = 1$), so

$$d(n) = x(n) - \hat{x}(n-1) \qquad (7.5)$$

The quantizer has only two levels (1 bit) and the step size is fixed. Each estimate of the signal $\hat{x}(n)$ differs from the previous estimate $\hat{x}(n-1)$ by only the step size δ. This method is used for signals with high sample-to-sample correlation, like highly oversampled speech waveforms. The coded output waveform is coded to 1 bit per sample with this simple quantizer. Linear delta modulation can use a quantizer with more than two levels, and the predictors can be of any order and need not be fixed to 1. A block diagram of a delta modulation system with 1-bit quantization and single-order prediction is shown in Figure 7.5, and Figure 7.6 illustrates the coding of a waveform.

A single-bit codeword specifies if the next sample of the signal is greater than or less than the previous sample. A "1" designates that

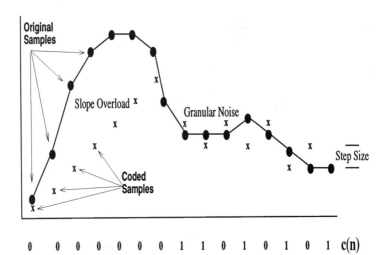

FIGURE 7.6
Delta modulation (Two types of quantization noise).

the next sample is greater than the last, while a "0" represents that the next sample is less than the last. If the sample is determined to be greater than the last, then the next sample is represented by the last signal plus a fixed increment; conversely, when the next sample is determined to be less than the last, it is represented by the previous sample minus the same fixed increment. This fixed increment is represented by the symbol δ. Both the sampling rate and the step size δ need to be chosen properly for delta modulation to be effective.

The coding error in delta modulation can be classified in two groups as shown in Figure 7.6. Slope overload occurs when the step size δ is not large enough to handle large sample-to-sample changes in the speech waveform. Granular noise occurs because the step size, δ, is too large to accurately narrow in on the speech waveform. Increasing δ would reduce slope overload distortion but increase granular noise. Conversely, decreasing δ reduces granular noise but increases errors due to slope overload. To reduce slope overload without affecting granular noise it is necessary to increase the sampling rate. For this reason delta modulation is often used on greatly oversampled signals.

7.4 Adaptive Quantization

The main tradeoff in signal quantization is making the quantization step size large enough to accommodate the maximum peak-to-peak range of the signal while keeping this step size small enough to minimize quantization noise. As discussed previously, nonlinear quantization addresses this problem in one manner. Another approach adapts the properties of the quantizer to the signal by having large quantizer step sizes in regions of the signal where the peak-to-peak range is high, and small step sizes when the peak-to-peak range is small, that is, to let the step size vary so that it matches the short-term variance of the input signal. Adaptive quantization schemes reduce the quantization error below that of μ-law quantization.

7.4.1 Adaptive Delta Modulation

Linear delta modulation can be modified so that the step size varies to better match the variance of the difference signal. The step size is increased or decreased when the output quantization code meets a predetermined criteria. An example of step size logic is as follows:

- Increase the step size by a multiplicative factor, $P > 1$, if the present code word $c(n)$ is the same as the previous code word $c(n - 1)$, otherwise, decrease the multiplier by a multiplicative factor, $Q < 1$.

This adaptation strategy is motivated by the bit patterns observed in linear delta modulation. Referring to Figure 7.6, it can be seen that the periods of slope overload are denoted by consecutive zeros or ones. Increasing the step size in these regions will reduce the slope overload. Periods of granularity are signaled by alternating codewords, and decreasing the step size minimizes quantization error in these regions.

7.4.2 Adaptive Differential Pulse Code Modulation (AD-PCM)

When an adaptive step size is introduced into DPCM systems, the new quantization scheme is classified as an adaptive DPCM (ADPCM) system. Because the signal being quantized is a difference signal, AD-PCM systems have predictors of first or higher order.

A simple but useful ADPCM system was introduced by Cummiskey, Jayant, and Flanagan in [25, 78]. The coder makes instantaneous exponential changes of quantizer step size, includes a simple first-order predictor, and has an adaptation strategy that depends only on the previous code-word. Figure 7.7 shows a block diagram of the coder. The signal is coded as follows:

$$\delta(n) = x(n) - \tilde{x}(n) \tag{7.6}$$

The difference signal, $\delta(n)$, is calculated and uniformly quantized into $\hat{\delta}(n)$ with step size $\sigma(n)$:

$$\hat{\delta}(n) = Q_{\sigma(n)}[\delta(n)] \tag{7.7}$$

The $\hat{\delta}(n)$ value is encoded as the digital output, $c(n)$. The step size $\sigma(n)$ is then scaled by the multiplier, $M_{c(n)}$, corresponding to $c(n)$, to adapt the step size for the quantization of the next sample:

$$\sigma(n+1) = M_{c(n)}\sigma(n) \tag{7.8}$$

The multiplier, $M_{c(n)}$, is selected by the "LOGIC" box in the diagram, where each code word corresponds to a different multiplier scaling of $\sigma(n)$.

Figure 7.8 shows the quantizer levels for a 3-bit coder. The difference signal, $\delta(n)$, is quantized to 0.5σ for $0 < \delta \le \sigma$, and $\delta(n)$ is quantized to 1.5σ for $\sigma < \delta \le 2\sigma$, etc. For $\hat{\delta}$ quantized as 0.5σ, the output codeword, $c(n)$, is set to 100, and the corresponding multiplier is M_{00}. (The value of the multipliers is not indicated in Figure 7.8. Multiplier symbols are displayed to show the tied correspondence to $c(n)$.) The most significant bit indicates the sign of the coded output. Note that there are only four distinct multipliers because the sign of quantizer output (the most significant bit) is not utilized in the adaptation logic. The adaptation depends only on the magnitude of $\hat{\delta}$.

The low-level multipliers, such as M_{00}, are kept small ($M_{00} < 1$) so that the quantizer is decreased if the step size is not small enough. Conversely, high-level multipliers, such as M_{11} in Figure 7.8, are fashioned to be large so that the quantizer step size is increased when it is found to be too small. The middle multipliers, such as M_{01} and M_{10}, are kept close to 1 so as not to change the size of the quantization step size when the difference signal is within the dynamic range of the quantizer.

The decoder works inversely. The adaptation size is adjusted by the multiplier associated with the encoded value. The value is decoded and

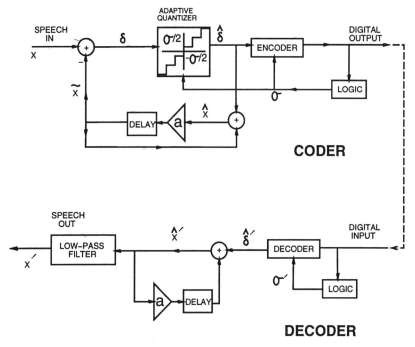

FIGURE 7.7
Adaptive quantization in differential PCM coding of speech
with a first-order predictor.

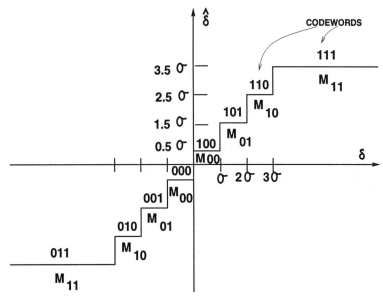

FIGURE 7.8
Quantization for an adaptive differential PCM speech coder.

$\hat{\delta}'(n)$ (the difference signal) is calculated; $\hat{x}'(n)$ is then calculated by summing $\hat{\delta}'(n)$ with $\alpha\hat{x}'(n-1)$.

$$\hat{x}'(n) = \hat{\delta}'(n) + a\hat{x}'(n-1) \qquad (7.9)$$

The output speech is produced by lowpass filtering $\hat{x}'(n)$ to remove the abrupt edges of the quantized steps.

ITU Recommendation G.726, ADPCM at 16, 24, 32, 40 kbit/s

In a more general ADPCM system, both the predictor and the quantizer adapt to the input signal. The ITU-T Recommendation G.726 [188] provides a standard for converting 64 kbit/s A-law or μ-law PCM data to 40, 32, 24, or 16 kbit/s ADPCM representation. The 32 kbit/s rate is the primary voice mode. The 24 and 16 kbit/s modes are for reduced capacity channels. The principal application of the 40 kbit/s rate is for encoding data modem signals.

The algorithm incorporates an adaptive quantizer, and an adaptive pole-zero predictor. The adaptive predictor uses 2 poles and 6 zeros. The coefficients of the pole-zero predictor are updated based on the input signal. For subjective evaluations of the output speech quality, the coder achieves a mean opinion score (MOS) of over 4; and as such, offers high-quality speech coding. (See Chapter 8 for an interpretation of MOS results.)

7.5 Vector Quantization

Vector quantization (VQ) [107] is a general class of methods that encode groups of data rather than individual samples of data. The idea is to exploit the relation among elements in the group to represent the group as a whole more efficiently than each element by itself. VQ systems usually operate on a parameter representation of speech as opposed to groups of time samples. VQ is a central component in most speech coding systems. It is frequently applied to quantize and code vocal tract information, often in the form of line spectral frequencies (LSFs). It is used to represent the excitation signal in code excited linear prediction (CELP) coders. These and other applications will be discussed in later chapters.

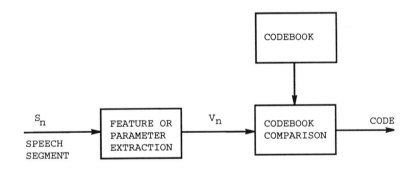

FIGURE 7.9
Vector quantization encoder.

Figures 7.9 and 7.10 show a block diagram of a simple vector quantization system. To encode speech data, a speech segment, S_n, is fed into a parameter extraction algorithm (such as linear predictive coding, see Chapter 4). The speech segment is referred to as a *frame* and is usually 10 to 25 ms long. The parameters vectors, V_n (e.g., LSFs), are compared

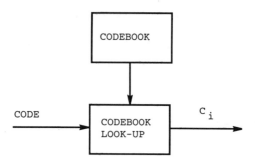

FIGURE 7.10
Vector quantization decoder.

with each vector, C_i, in a codebook using a distance metric to determine which codebook vector best matches the input vector. The input vector and the codebook vectors have the same number of elements. The index of the closest match is stored or transmitted as a single code word for each speech segment. The codebook is precalculated and is stored exactly in the decoder as it is in the encoder.

To decode, the code is sent through a codebook lookup process. The transmitted or stored codeword is an index into the codebook. This index is the same as determined during the coding process. Based on the index, the vector C_i is retrieved. This vector is determined, by the codebook generation process, to best represent vectors similar to the original vector V_n. The vector C_i is further processed to produce synthesized speech, depending on the information it contains. For a complete speech coder, a synthesis algorithm would incorporate additional information (such as pitch and voicing information) to reconstruct the speech segment.

Although there are many different algorithms for creating a codebook, they all perform the same basic tasks. For L codebook entries, the M-dimensional vector space is sectioned into L nonoverlapping cells. This sectioning is usually performed based on a set of example speech vectors referred to as training vectors. In many implementations, C_i is the centroid of the training vectors within cell i. The centroid is the multidimensional mean of those training vectors for a particular cell.

The centroids of the cells represent the output code vectors associated with the corresponding cells. In other words, during the encoding process, when an input vector falls within a particular cell, the index of that cell will be transmitted as the codeword. For the decoding process, the centroid of the cell will be the output vector. Figure 7.11 displays a two-dimensional vector space, partitioned by cell boundaries, with the centroids marked. The cells are numbered with a k-bit codeword where $k = \log_2 L$. The dimensions of vector space are v_{n1} and v_{n2}, the first and second elements of vector V_n.

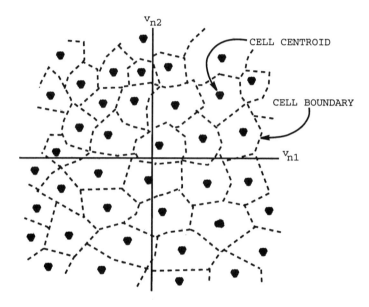

FIGURE 7.11
Vector quantization partitioning of a two-dimensional vector space; centroids marked as dots.

One simple method for creating the codebook is to partition the M-dimensional vector space into k uniform hypercubes. More efficient techniques dedicate more cells to regions of the vector space that are more densely populated by the training speech vectors.

7.5.1   Distortion Measures

A *distortion measure* indicates how similar two vectors are. It is used to decide how close an input vector is to a codebook vector and is also used in the training of the codebook.

A common distortion measure is the sum of the squared differences. It is computed as:

$$\text{Squared Error}(V_n, C_i) = \sum_{j=0}^{M-1} (v_{nj} - c_{ij})^2 \qquad (7.10)$$

where v_{nj} is the jth element of vector V_n, and c_{ij} is the jth element of vector C_i. In this case, all differences between vector elements are weighted equally.

The distortion can be adjusted to weigh the difference between certain vector elements more than others. The weighted error is:

$$\text{Weighted Error}(V_n, C_i) = \sum_{j=0}^{M-1} [w_j(v_{nj} - c_{ij})]^2 \qquad (7.11)$$

If the variance of the vector element v_{nj} is different from that of v_{nk}, and differences relative to the respective variances are important, the weighting can be used to normalize by the standard deviation, σ_j. The σ_j is estimated from the training data set as the square root of the variance of element j. The weighting is then $w_j = 1/\sigma_j$. In this case, differences are treated inversely proportional to the variance of the element in the training set.

For the coding of LSFs, a perceptually motivated weighting was suggested by Paliwal and Atal in [129]. The metric incorporates two weightings for a tenth-order LSF vector as:

$$d(F_i, F_k) = \sum_{j=1}^{10} [w_j b_j(f_{ij} - f_{kj})]^2 \qquad (7.12)$$

where F_i and F_k are vectors of LSFs; f_{ij} is the jth frequency of F_i; and the weights are defined as:

$$b_j = \begin{cases} 1.0, \text{ for } 1 \leq j \leq 8 \\ 0.8 \text{ for } j = 9 \\ 0.4 \text{ for } j = 10 \end{cases} \qquad (7.13)$$

$$w_j = (P(f_j))^r \qquad (7.14)$$

where $P(f_j)$ is the LPC power spectrum at the frequency, f_j, and r is an experimentally determined constant set to 0.15.

The weighting w_j is perceptually based because it weights the distortion more heavily for frequencies with more spectral power. This corresponds to treating the regions around formants more importantly than other areas, and high energy formants more than low energy formants. The b_j term discounts the high frequency LSFs because the lower frequency portions of the spectrum are perceptually more significant.

7.5.2 Codebook Training

Codebook sizes in use range from 8 bits (256 entries) or smaller up to about 12 bits (4096 entries). (These numbers apply to single-stage codebooks. The following section discusses alternate approaches for more accurate quantizations using more bits.) The best codebook for a given set of training data is one which will minimize the quantization error over the vectors to be quantized. However, even for moderately sized codebooks, an exhaustive search for the best codebook results in an impractically large number of computations. So, codebooks are generated, or trained, using statistical clustering algorithms.

The LBG algorithm, named after Linde, Buzo and Gray [103], is a widely used clustering method for generating a codebook. The algorithm begins with an initial set of codebook vectors, assigns the training vectors to the nearest codebook vector, then recomputes the codebook vector as the centroid of all the vectors assigned to that codebook entry. The process is repeated until convergence, or no reduction in the overall quantization error. The steps are listed in more detail below.

1. Randomly select L vectors from the training data to initialize the codebook.

2. Determine which codebook entry, C_i, is closest to the training vector, V_n, by using an appropriate distortion measure. Assign training vector, V_n, to the closest codebook entry. Repeat for all V_n (over all n) in the training data set.

3. Calculate a new codebook entry, C_i^{new}, based on the centroid of the training vectors assigned to the current codebook entry, C_i. Iterate for all C_i (over all i). The new codebook becomes the current codebook.

4. Repeat 2 and 3 until the codebook converges such that Distortion(C_i^{new}, C_i) is less than a low threshold value for all C_i (over all i).

The convergence criteria is based on a threshold because of the possibility that a few training vectors may oscillate assignments between two codebook entries with each iteration. In that case, each iteration changes the codebook vectors slightly numerically, but does not alter or improve the codebook performance. The algorithm yields a local minimum in the effort to produce an optimized codebook with the lowest quantization error, but not necessarily a global minimum.

7.5.3   Complexity Reduction Approaches

For a given VQ arrangement, it is necessary to increase the size of the codebook to reduce the quantization error. However, as the number of codebook entries grows to 20 or 25 bits ($2^{25} = 33,554,432$) or more, a full search of the codebook to quantize each input vector becomes impractical. Different approaches have been developed to address this issue. Split VQ (SVQ) and multi-stage VQ (MSVQ) are two examples of *product codes*. In a product code, the quantization of the vector is distributed among multiple codebooks, and the results from each are combined for the overall quantization. A product code has as many effective codebook entries as the product of the number of entries in each component codebook, but the search effort is the sum of the search efforts for each component codebook. Also, the total storage space is only the sum of the component codebooks.

During the training process, tree-structured VQ (TSVQ) overlays the codebook onto a tree structure. This allows the search process to follow branches of a tree structure, based on decisions at the nodes, to reach the closest codeword. TSVQ yields very quick search times.

Both product codes and tree structures reduce the search effort at the expense of decreased quantization performance relative to a full-search, single codebook with the same number of bits.

Split Vector Quantization

For split VQ [129, 130], the input vector is segmented into multiple parts. Each part is vector quantized individually, and the code indices are transmitted or stored. At the decoder, the component indices are used to look up the quantized values from the corresponding component codebooks. Finally, the components are concatenated to produce the output quantized vector.

For example, [129] reported on the SVQ of LSFs at 24 bits/frame. The LSF vector was split into 2 parts. One part included LSFs 1 through 4; and the second, 5 through 10. Each split was vector quantized with 12 bits for a total of 24 bits. This uneven split gives more quantization accuracy to the perceptually more important lower LSFs. The quantization performance, as measure by the spectral distortion, of the 24 bit SVQ was better than a 32 bit scalar quantizer.

Multi-Stage Vector Quantization

Multi-stage VQ [101], also known as *cascaded VQ*, uses a sequence of vector quantizers, each operating on the output of the previous stage. The first stage quantizes the input vector. The second stage quantizes the error between the input vector and the quantized vector from the first stage. This cascading is repeated for each stage. Four-stage VQ coders are commonly used. The final output quantized vector is the sum of the outputs from each stage.

Because the impact of a future stage of quantization on the overall distortion is not known at the current stage, multiple close match candidates are carried along to the next stage. This is referred to as the "M-best" search procedure [101], where M candidates with the lowest distortion from the current stage are carried to the next stage for consideration. This process is repeated at each stage. At the final stage, the coding with the lowest overall distortion (all stages) is selected from the M final candidates. Experimentally, setting M equal to 8 has been found to offer good overall performance.

In the Federal Standard mixed excitation linear prediction (MELP) coder of [156], LSF vectors are vector quantized with a 4-stage MSVQ. The first stage uses 7 bits, and each of the three succeeding stages uses 6 bits for a total quantization of 25 bits.

Tree-Structured Vector Quantization

Tree-structured VQ systems [107, 138] are designed for quick search times. The TSVQ maps the codebook onto a tree structure during the training of the codebook. Most often, the tree is a binary structure.

One method to generate a TSVQ begins by randomly choosing two training vectors. The remaining training vectors are assigned to one or the other, whichever is closer; and the centroids are computed for the two clusters. Each of the two clusters is further split into two clusters in the same manner. Various stopping criteria can be devised for deciding when to terminate the splitting process, such as examining the variance of the cluster. The centroid of the cluster at the end of a branch, or terminal node, becomes the codeword.

To search for the closest codeword, the input vector is compared to the two cluster centroids that form the split at the current branch. The branch is followed that corresponds to the centroid that is closer to the input. This is repeated until reaching a terminal node. This search requires only the number of code vector comparisons required to reach a final node. This number is variable based on the number of branches, but is only a small fraction of the total number of codewords as required for the full codebook search method.

To generate a more optimal codebook, the splitting of clusters can continue until there is a large number of terminal nodes (many more than the final number of codebook entries). The terminal nodes are then regrouped (combining two adjacent branches back together) based on which recombinations increase the overall distortion the least. This regrouping is repeated until the the number of terminal nodes is equal to the desired codebook size [138].

Because the tree structure method doubles the space required to store the codebook, practical implementations for large codebook sizes incorporate the split VQ method. In this case, each segment of the vector is quantized with a TSVQ and the resulting vector segments are concatenated to form the overall output vector. In [21], Collura and Tremain present results of accurate quantization of LSFs using 25 to 27 bits/frame with tree-structured 2- and 3-split vector quantizers. For the 2-split, the 4 lower and 6 upper LSFs were quantized separately; and for the split into three sections, the 3 lower, 3 middle, and 4 upper were grouped.

7.5.4   Predictive Vector Quantization

By the nature of vector quantization (quantizing all elements at once), dependencies between elements in the same vector are accounted for and contribute to efficient quantization. Between-frame dependencies (across time) are not incorporated, however, in standard, single-frame quantization. Predictive VQ (PVQ) [173] attempts to reduce interframe redundancy to increase coding efficiency by predicting the current input vector based on the previous vector. The residual, or error, from the prediction is fed to a vector quantizer. PVQ is the vector extension of predictive quantization as described in Section 7.3.1.

Figure 7.12 displays a diagram of a predictive VQ system. The predicted vector is subtracted from input vector. The predicted vector is based on the quantized vector, C_i, from the previous input, V_{n-1}. The difference is passed on to the VQ for quantization.

Both the codebook and predictor are trained at the same time. In

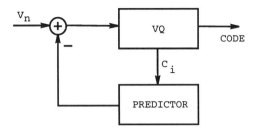

FIGURE 7.12
Block diagram of predictive vector quantizer.

most cases, an iterative method is employed where the predictor is trained given the codebook; then, the codebook is trained with the newly derived predictor. The process is repeated to refine both the predictor and the codebook.

Advanced approaches to LSF quantization is an area of significant current research. Recent methods have found it advantageous to switch the predictor and quantizer between two different modes. The general method is referred to as *switched-predictive VQ* [173]. The reason for the mode switching is that speech is, at times, slowly varying, and so prediction across frames works well. At other times, speech changes rapidly, and a strong predictor actually reduces performance. For the transient portions of speech, either no prediction [174] or a predictor trained on the rapidly varying sections [68] is used. The mode that gives better quantization (lower distortion) is selected, and one bit is transmitted to indicate the mode.

In [174], Zarrinkoub and Mermelstein use a first-order prediction when the speech spectrum changes slowly. For rapid spectrum changes, they used a quantizer trained on the that type of speech, without a predictor. Given the mode, rapid or slow, a 2-split VQ is used to code the LSFs. For quantizing 10 ms frames, they report 1.05 dB average spectral distortion with 2.9% outliers having distortion greater than 2 dB. The 10 ms quantizer uses 9 bits for each split and one mode bit, for a total of 19 bits/frame. The setup is the same for 20 ms frames, but the performance is 1.13 dB average distortion with 3.6% outliers and a 21 bit/frame overall rate (10 bits first split, 10 bits second split, 1 bit mode).

Heinen, Adrat, Steil, Vary, and Xu reported on a 22 bit/frame LSF quantizer in [68]. The system uses two predictors, one trained on stationary speech, the other on transient. The quantizer is a 2-stage split VQ. The first stage quantizes the prediction error with 9 bits over all the 10 LSFs as one vector. The second stage splits the error from the first stage into two 5-element vectors, and quantizes each split with 6 bits. The overall rate is broken down as 1 bit to pick the quantizer, 9 bits for the first stage, and (6+6) bits for the splits of the second stage. They reported 1.2 to 1.4 dB average spectral distortion.

A switched predictive VQ system was recently reported in [116] by Mc-Cree and De Martin. For a 21 bits/frame quantizer operating on 20 ms frames, the average spectral distortion was 0.97 dB, with an outlier rate of only 0.81% (distortion > 2 dB). The system employs two predictors paired with two 4-stage, 20 bit codebooks to code the prediction residual. The additional bit selects the better performing predictor/codebook pair.

Chapter 8

Quality Evaluation

Evaluating voice coder quality is a difficult task. No simple formula or mathematical calculation can provide a score indicating the quality of the decoded output speech. The problem lies in the fact that speech quality is inherently tied to speech perception. The perceived quality and understandability of coded speech depends on numerous conditions including speech content, speaker individuality, background noise, coding channel losses, and the listener. Removing some of these variables and averaging over the others lessens the scope of the discrepancies; however, different listeners will disagree on the quality of a single example of decoded speech.

The signal-to-noise ratio (SNR) can be computed easily. It measures the difference between the original speech and the decoded output speech. Although SNR is a good objective basis to compare the quality of waveform coders, it does not provide a useful measure for many speech coders. Many speech coders do not attempt to mimic the original waveform; rather, they attempt to mimic the perceived sound. These coders extract the perceptually significant parameters from the input signal, and use the parameters to reconstruct the signal. For these coders, the error between the original and synthesized waveforms might be quite high, while the perceived differences in the sound might be low.

Because all speech coders utilize perceptual qualities to some extent, subjective quality measures based on listening tests are usually more relevant than objective measures such as the SNR. Current research efforts are being directed towards perceptually based objective measures.

8.1   Objective Measures

Objective measures offer ease of computation. The SNR gives an indication of how well the waveforms match, a useful figure for waveform coders. It depends on matching the phase of the synthesized waveform to the original. For speech coders which do not match the specific waveform, spectral measures provide a idea of the closeness of fit of the shape of the short-term spectrum.

8.1.1   Signal-to-Noise Ratio

The signal-to-noise ratio (SNR) computes the energy in the original signal relative to the energy of the noise, where the noise is the error between the original and synthesized.

The SNR in dB is:

$$SNR = 10 \log_{10} \frac{\sum_{n=0}^{N-1} s^2(n)}{\sum_{n=0}^{N-1} (s(n) - s'(n))^2} \qquad (8.1)$$

where $s(n)$ is the original signal; $s'(n)$ is the synthesized, decoded output speech; and N is the length of the speech segment being measured.

The SNR weighs high amplitude portions of the signal more than low amplitude portions. Low amplitude segments can be perceptually important, and small errors, proportional to the low amplitude, can result in noticeable degradations. The segmental SNR addresses this concern by computing the SNR as the average over a number of short segments. Because the error is computed relative to the signal energy for short segments, small amplitude segments contribute equally to the overall measure.

The segmental SNR is computed as:

$$SNR = \frac{10}{M} \sum_{m=0}^{M-1} \log_{10} \left(\frac{\sum_{l=0}^{L-1} s^2(mL+l)}{\sum_{l=0}^{L-1} (s(mL+l) - s'(mL+l))^2} \right) \qquad (8.2)$$

where s is the original signal; s' is the synthesized, decoded output speech; and ML is the length of the speech. The inner sum is over the short speech segments, of length L, typically 10 to 20 ms.

8.1.2 Spectral Distance

Spectral distance measures are more appropriate for vocoders (that are not waveform coders), because spectral distance does not depend on the phase of the synthesized signal relative to the original. Signals with the same frequency content, but different short-term phase, can sound quite similar. However, spectral distance measures generally are not able to capture coding degradations due to transient, temporal discontinuities. These types of degradations are obvious to the human ear.

The log spectral distance integrates the differences in the short-term magnitude spectrum of the original and coded speech segments. It is defined as:

$$\text{DISTANCE}_{\log} = \int_{-\pi}^{\pi} |\log |S(\omega)|^2 - \log |S'(\omega)|^2 |d\omega \qquad (8.3)$$

where $S(\omega)$ and $S'(\omega)$ are the Fourier transforms of the original and synthesized speech, respectively.

For easier computation, the discrete Fourier transform (DFT) of $s(n)$ and $s'(n)$ can be substituted for the Fourier transforms of the previous equation, and a sum replaces the integral.

Reference [131] suggests Euclidean distance measures for linear prediction (LP) representations of the speech spectrum, including reflection coefficients and log area ratios. The distance is computed as:

$$\text{DISTANCE}_{\text{Euclidean}} = \sqrt{\sum_{i=1}^{p}(k_{Si} - k_{S'i})^2} \qquad (8.4)$$

where k_{Si} and $k_{S'i}$ are the ith coefficient of the original and synthesized LP analysis, and p is the order of the LP analysis.

8.2 Subjective Measures

The objective measures discussed in the previous section do not account for how the decoded speech signal is perceived. Because most speech coders do not quantize the time-domain signal directly, it is difficult to algorithmically interpret the perceptual significance of coding degradation to a perceptually coded acoustic signal. To provide quantitative information to compare the quality of speech coding systems, it is necessary to listen to the output.

Current speech research efforts rely on subjective listening tests to both guide research decisions and select among competing coders. Intelligibility tests most often follow the Diagnostic Rhyme Test (DRT) paradigm. For tests that measure speech quality, the mean opinion score (MOS) is used most often.

8.2.1 Intelligibility

The most basic information that is preserved, or distorted, by the coding system is message content. An intelligibility test asks the listener to identify which words were spoken. The most widely used intelligibility test is the Diagnostic Rhyme Test (DRT).

Diagnostic Rhyme Test

The Diagnostic Rhyme Test [36, 163] asks the listener to distinguish between rhyming words that differ only in first consonant sound. The listener sees both written words and then hears one of the pair. The task is to select the appropriate word. Example word pairs might include:

- hit / fit

- moon / noon

- you / rue

- bid / did

For reliable test results, a large number of word pairs, speakers of both sexes, and listeners is required. The test result is reported as the percentage of correct responses with an adjustment for guesses. The range of possible values is from 0 to 100%, and is computed as:

$$\text{DRT} = \frac{Correct - Incorrect}{Total} * 100 \qquad (8.5)$$

Modified Rhyme Test

The Modified Rhyme Test (MRT) was suggested by House et al. in [75]. It broadens the structure of the test setup. The MRT is a multiple-choice test where the listener selects from six similar sounding words for each audio example of the speech coder. For half the test, the words differ only in the first consonant. For the other half, they differ only in the last consonant. The MRT is not widely used.

8.2.2 Quality

While intelligibility tests measure whether the correct phonemes can be perceived from the coded speech, intelligibility testing does not account for how the speech "sounds." It is possible for coders to produce speech that might be described as "mechanical," "raspy," "buzzy," or "thin," but at the same time, be highly intelligible. Quality testing attempts to rate how good the speech sounds relative to the presence or absence of degrading coding artifacts. Quality testing is inherently subjective, and the results vary significantly with different speakers and listeners.

Diagnostic Acceptability Measure

The Diagnostic Acceptability Measure (DAM) [162] is an extensive, systematic method to rate the quality of coded speech. The DAM method employs trained listeners who are frequently retested with controlled inputs of known quality to normalize for individual preferences. The listeners grade the speech along individual scales for a number of qualities including "rasping," "muffled," and "fluttering." The listeners also rate the overall acceptability.

Mean Opinion Score

The Mean Opinion Score (MOS) is a commonly used quality test where the listener rates a coded phrase based on a fixed scale [191]. The scale ranges from 1 to 5. For each level, an accompanying word describes the quality of speech. The following scores describe the quality of a speech signal.

- 5 = Excellent

- 4 = Good

- 3 = Fair

- 2 = Poor

- 1 = Unacceptable

Listeners are presented utterances one at a time and asked for their opinion as to the quality of the speech in the terms listed above. This test is performed for each utterance over many listeners. Listener group size ranges from 15 or 20 to over 40. As with any test of this nature, a larger listener group provides more reliable results. The mean of the opinion scores for each utterance is calculated and recorded as the MOS of the utterance. To rate a speech coding system, the test includes a large number of utterances.

Original, uncoded speech should score a 5. Most speech coding systems score between 3 and 4. A coding system that scores above 4 provides very high quality.

MOS test results can suffer from significant variability. Opinions on the quality of some speech utterances vary greatly with different listeners and even for the same listener on a different day. Results cannot be compared from different tests and listener groups with absolute precision.

Reference [158] presents results from a comparative study of MOS testing versus DAM testing for quality assessment. The study included a number of speech coders covering a wide range of quality. The study concluded that if sufficient care is taken in the structuring and presentation of the MOS test, the MOS test can produce reliability and resolution equivalent to the DAM test.

Degradation Mean Opinion Score

Listener responses in subjective listening tests are influenced by a number of sources of variation, e.g., speech material, speaker voice characteristics, presentation order, time effects. Unless controlled in some way, these variables can bias the outcome of the experiment. The Degradation Mean Opinion Score (DMOS) testing procedure was developed to reduce these biases [191]. The DMOS method is also referred to as Difference MOS.

Each test utterance in a DMOS test is preceded by the original reference utterance. The listener is asked to rate the degradation of the test utterance as compared to the original, undistorted utterance. The listener rates the reduction in quality on a scale from 1 to 5. On this scale, a 1 corresponds to much worse than the original, 4 is the same as the original, and 5 means better than the original. The process of comparing the test utterance with the reference stabilizes the test results. The DMOS paradigm is particularly useful for rating coder quality for speech severely degraded by background noise or transmission channel errors. An example of the application of DMOS testing is presented in Section 12.2.3 in the context of perceptual speech coding.

Pair-Wise Comparison

A pair-wise comparison, also called A/B comparison, rates speech processed by two different coding schemes. The original speech is processed by both algorithms and presented to a listener. The listener selects the one with better perceived quality. The test is repeated for several speakers, many utterances, and, ideally, many listeners.

Pair-wise comparisons are easy to organize and reasonably reliable. The nature of the test requires less training and calibration of the listeners.

8.2.3   Background Noise and Channel Conditions

During listening tests, the test setup conditions refer to the influences on the coded speech other than speaker and phrasing. The test conditions are designed to simulate the type of real-world difficulties under which the coder might be employed. The most obvious test condition is clean speech, or no background noise. However, the intended use of many speech coders requires them to operate in noisy environments. Examples include car noise for cellular telephone applications, and aircraft cockpit background for military applications. Because the noise might be very different from any speech sounds, a coder that attempts to fit the noise to a speech model might produce annoying or distracting synthesized output. In other words, the noise sounds nothing like it did originally. How well the coder reconstructs the background noise can greatly influence the overall perceived quality.

Quality tests under different channel conditions are important for coder applications in radio systems including satellite and cellular. The test condition simulates a certain level of transmission errors. The errors are usually assumed to be randomly distributed in the coded data, for example a 1% random bit error. The manner in which the coder organizes the bitstream of coded speech parameters, and any additional error protection coding, greatly impacts how the quality of the output speech will fare under these tests.

8.3 Perceptual Objective Measures

The objective measures discussed in Section 8.1 are easily computed on synthesized coded speech, but have no direct relation to the perceived quality of the speech. The subjective listening tests of Section 8.2 measure the perceived quality, but at significant expense due to the number of listers and time required. Current research efforts are aimed at developing objective measures that correlate with listening-test quality ratings. The idea is to have an automated processing algorithm that can predict the MOS rating of a listening test. While these efforts show promise, they have not eliminated the need for formal subjective listening tests for coder evaluation.

The International Telecommunications Union (ITU) standardized an objective Perceptual Speech Quality Measure (PSQM) [8] under the recommendation P.861 [190], originally in 1996 and with later revisions. The system maps the original and coded/decoded synthetic version into a perceptual frequency representation. The frequency representation is based on the Bark spectral representation (discussed in Chapter 6). Time and frequency masking is taken into consideration, as well as the nonlinear perceptual transformation of signal power levels into perceptual power levels. After both the original and coded versions are transformed, they are compared in the perceptual domain. The level of difference in the perceptual domain is then mapped to a MOS number based on an experimentally derived mapping function.

Another objective perceptual quality measure is the Bark Spectral Distortion (BSD) [167] and its further refinements [169, 170, 124]. The BSD is the average squared distance between the perceptual loudness of the original and coded speech, that is, the squared difference between Bark spectra. To obtain the perceptual loudness of a signal, the speech is transformed into the critical-band frequency domain, and the intensities are adjusted nonlinearly in a perceptual manner [167].

The Modified Bark Spectral Distortion (MBSD) [169] includes an explicit noise-masking threshold (simultaneous masking as discussed in Section 6.4.1) to decide when distortions are large enough to be included in the overall distortion calculation. Distortions falling below the masking threshold are not included. Reference [170] improves the correlation of the MBSD with listening test results by mapping the MBSD to DMOS numbers instead of MOS numbers. In [124], Novorita incorporates temporal masking (discussed in Section 6.4.2), both forward and backward, into the BSD measure to improve the correlation with subjective listening tests.

Chapter 9

Voice Coding Concepts

This chapter describes several basic vocoders, in part, for historical background, but primarily for the purpose of introducing basic concepts that have been incorporated into subsequent, more complex coders.

In 1939, Homer Dudley published a description of the first vocoder [33] in the *Bell Labs Record*. The term *vocoder* was derived from VOice CODER. The vocoder was conceived for efficient transmission of speech signals over expensive long-distance telephone circuits. Vocoders compress the speech signal and have evolved to become more efficient over the years. Dudley is credited with being the first to show that speech signals can be transmitted over a fraction of the bandwidth occupied by the original signal when properly coded. Dudley's vocoder, a type of channel vocoder, was the first device to realize the promised economy [143]. The fact that increased economy can be achieved with a vocoder implied that much of the actual speech signal is redundant.

Many vocoders are based on the source-filter speech model (see Section 2.3). This approach models the vocal tract as a slowly varying linear filter. The filter is excited by either glottal pulses (modeled as a periodic signal), turbulence (modeled as white noise), or a combination of the two. Similarities exist between the source-filter model and actual speech articulation. The source excitation represents the stream of air blown through the vocal tract, and the linear filter models the vocal tract.

Unlike waveform coders, which attempt to reconstruct accurate representations of the time-domain waveform, vocoders reproduce a signal that is perceptually similar. It is well established that the human auditory system performs a short-time frequency transform on acoustic signals prior to neural transduction and perception [38]. Exact preservation of the time waveform is not necessary for perceptually accurate signal representation.

Although early vocoders such as Dudley's had the aggravating side effect of sounding unnatural, modern vocoders can sound surprisingly natural and, in some cases, give insight into speech enhancement methods. This stems from the fact that current vocoders have incorporated many of the properties of the acoustic theory of speech production. That is, these vocoders utilize the properties of the vocal tract to analyze and synthesize speech. Speech enhancement is the process of making speech sound perceptually better, which is often performed by reducing noise in the speech signal.

Efficient vocoders are useful for many speech processing applications, including data compression for transmission and storage, and for secure transmission of speech signals. High speech intelligibility is possible at much lower bit rates than is possible by direct coding of the speech waveform (64 kbits/sec is the coding rate of standard μ-law coding used in present day telephony). Vocoders also perform a transformation into the frequency domain, convenient for other types of speech processing. Manipulation of the data in this domain facilitates many speech processing functions, such as speaker transformation (changing one person's voice to sound like another's), speech enhancement, or time-scale modification of speech (changing the rate of speech without altering the perceived frequency characteristics).

The following discussion concerns channel vocoders, formant vocoders, and linear predictive coding (LPC) vocoders. In all of these vocoders, speech is analyzed in overlapping time segments, each of which is treated as the response of a linear system to an excitation signal. Further, the excitation is assumed to be made up of either a periodic impulse train (possibly modified to more closely resemble a glottal pulse train) or random noise. For each time segment of speech, the excitation parameters, and the parameters of the linear system are determined and then used to synthesize the speech during reconstruction.

This chapter also presents the concept of sinusoidal modeling for speech coding. Sinusoidal modeling is an appropriate choice for coding periodic signals, speech, and general audio signals. The successful Multi-Band Excitation (MBE) coder of Section 11.1 can be considered to be a type of sinusoidal coder.

9.1 Channel Vocoder

The concepts demonstrated in the channel vocoder are relevant for the subsequent discussion of the perceptual speech coder in Chapter 12. The perceptual speech coder uses subband coding, the process of dividing the frequency spectrum into many channels and coding the output of these channels.

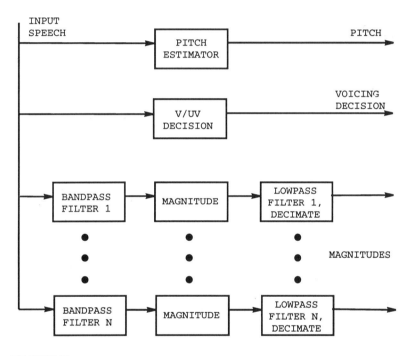

FIGURE 9.1
Channel vocoder analysis of input speech [137].

Figure 9.1 is a diagram of the analysis portion of the channel vocoder [137, 131]. Channel vocoders analyze the speech in the frequency domain by estimating the energy in discrete frequency bands (channels) covering the range of frequencies below half the sampling rate. The processing involves bandpass filtering (yields multiple bandlimited channels), rectification (magnitude or absolute value operation), and lowpass filtering (envelope detection), and finally decimation, so that the magnitude signal is represented at a reduced sample rate [137]. This sequence of operations approximates the magnitude of the short-term Fourier transform.

The speech is then classified for each time segment (frame) of speech as either voiced or unvoiced and synthesized accordingly in reconstruction. If voiced, the pitch period is estimated and stored or transmitted. The quantization of the output parameters (magnitude signals, voiced/unvoiced signals, and pitch signals) of the channel vocoder is often performed with two different methods, one for the excitation parameters and one for the magnitude signals. For segments labeled unvoiced, the excitation can be coded with only one bit and regenerated as random white noise. For voiced segments, not only must one bit be spent to declare the excitation voiced, but 7 to 10 bits are required to quantize the period of the excitation (the pitch) [38, 62]. This is often done with simple linear quantization. The magnitude signals are quantized on a log scale, or the log difference of two neighboring frequency channels is quantized.

The number of frequency channels is a design decision of the channel vocoder that represents a trade-off between bit rate and synthesized speech quality. The number of channels is fixed for a particular implementation. Typical implementations utilize from 15 to 20 channels, or bands, over the 0 to 4 kHz band. Channel spacing and bandwidth is usually nonlinear, with more bands of narrower bandwidth being used to cover the more perceptually significant lower frequencies.

Figure 9.2 shows the synthesis portion of the channel vocoder. The bit containing voicing classification information (voiced or unvoiced) determines the source signal. The source is scaled by the magnitudes, and the bank of bandpass filters limits each channel to the appropriate frequency band. The signals from each of the bands are added to realize the synthesized speech signal.

9.1.1 Implementations of the Channel Vocoder

It was noticed that the adjacent output samples of each channel in the channel vocoder were highly correlated. This led to the use of adaptive differential pulse code modulation (ADPCM) coders for the channel magnitudes [25, 52]. Other methods were also attempted to reduce the bit rates of the channel vocoders without significant perceptual degradation of the speech signal. One method reduced the number of channels

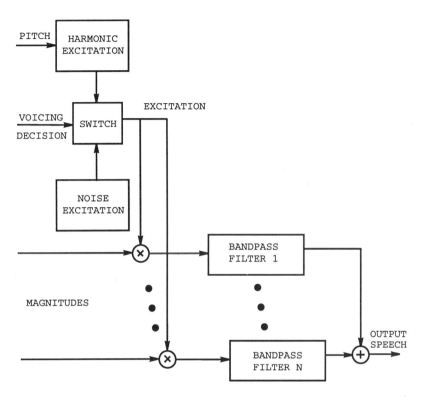

FIGURE 9.2
Channel vocoder synthesis of decoded output speech [137].

to be transmitted. Peterson and Cooper [133] suggested transmitting only those channel signals that are greater than their neighbors (that is, adjacent channels in frequency). They transmitted about 1/3 of the output samples of the channel vocoder, but only reduced the bit rate by about 30 percent due to the need to transmit side information to determine which channels were transmitted and which were deleted.

The Joint Speech Research Unit (JSRU) of the United Kingdom developed a practical channel vocoder operating at 2400 bit/s [74]. The coder splits the frequency band into 19 channels. It reduces the redundancy in the magnitude signals from those channels by coding the difference of the signals across the channels. One advantage of this vocoder is robustness to transmission errors. Because of the manner in which the spectral shape information is coded (difference of adjacent channels), bit errors only slightly affect the spectral shape and the resulting intelligibility.

9.2   Formant Vocoder

For most voiced speech sounds, there are several prominent maxima in the spectral envelopes. These represent the resonances of the vocal tract and are called *formants* (see Section 2.2.2). Adult speech is characterized by three formants in the frequency range below 3 kHz. The formants are plainly visible in the spectrogram of Figure 2.11 as the darker bands that vary smoothly in frequency, across time. The second formant (second lowest in frequency) carries perceptually significant information concerning phoneme identity.

Formant vocoders attempt to locate the formant frequencies in each frame, or short-time segment, of speech (typically 10 to 30 ms). These formant locations are transmitted and utilized in synthesis to characterize the filter portion of the source-filter model. The excitation signal is generated similarly to that of the channel vocoder with pitch information and a voiced/unvoiced decision. A schematic of a formant vocoder is shown in Figure 9.3 [143]. The formant vocoder can be viewed as a bit-rate saving extension to the channel vocoder. The formant vocoder transmits only a compact representation of the formants, instead of all the channel magnitudes in the channel vocoder. The filter portion of the formant vocoder is forced to always transmit the same amount of information, the perceptually important large spectral peaks. For vowels, excited by the vocal cords, the formant frequencies, amplitudes, and bandwidths suffice to specify the entire spectral envelope [37]. The spectral contribution from a single formant can be expressed in the z-domain as:

$$H(Z) = \frac{1 - 2e^{-BT}\cos(2\pi FT) + e^{-BT}}{1 - 2e^{-BT}cos(2\pi FT)z^{-1} + e^{-2BT}z^{-2}} \qquad (9.1)$$

where B is the bandwidth of the formant, T is the sampling period, and F is the formant location in frequency.

In practice, the formants are located by picking peaks from a representation of the short-time spectral envelope. Fourier and LPC spectra and cepstral representations have been used. The main problem with formant vocoders is that formant tracking (the process of picking the cor-

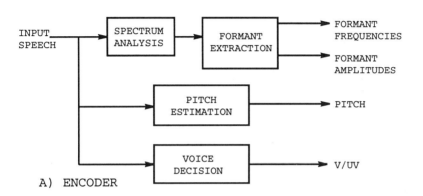

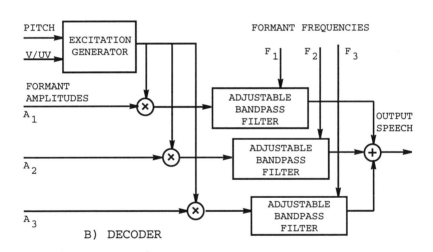

FIGURE 9.3
Formant vocoder analysis and synthesis [143].

rect formant frequencies for each consecutive time segment) is an inaccurate science, and errors in formant frequencies cause large degradations in the synthesized output speech. The difficulties in formant tracking are due to the fact that the peaks in the short-time spectrum do not always represent a formant frequency; in fact, in some speech segments, the formant frequencies will not even be clear as local maxima. In other words, while specifying the formant frequencies and bandwidths determines the spectral envelope exactly, the converse is not necessarily true. Differing formant group specifications can produce nearly the same spectrum. In [131], Papamichalis elaborates on the details associated with formant coders. Synthesizing speech from a formant representation has been researched more thoroughly in the context of direct synthesis of known text than speech coding.

9.3   The Sinusoidal Speech Coder

Another coder based on the source-filter model of speech production is the sinusoidal coder. As with most source-filter models, the way in which the excitation is coded is what sets this coder apart from the rest. In the sinusoidal coder, the excitation is assumed to be composed of sinusoidal components with specific amplitudes, frequencies, and phases [112].

9.3.1   The Sinusoidal Model

The sinusoidal model assumes the source-filter model of Section 2.3, where the source (model of the vocal cord glottal excitation) is modeled by a sum of sine waves. It was shown that within certain parameters, both voiced and unvoiced excitation can be modeled effectively in this way [113]. In short, voiced speech can be modeled as a sum of harmonic sine waves spaced at the frequency of the fundamental, with phases tied to the fundamental. Unvoiced speech can be represented as a sum of sinusoids with random phases.

The speech waveform can be modeled by:

$$s(n) = \sum_{l=1}^{L} A_l cos(\omega_l n + \phi_l) \qquad (9.2)$$

where A_l, ω_l, and ϕ_l represent the amplitude, frequency, and phase of each of the L sine wave components, respectively.

9.3.2 Sinusoidal Parameter Analysis

In the general case of sinusoidal modeling, the frequency location of the sinusoids is not constrained to the pitch harmonics. This approach, without quantization, accurately models signals containing multiple speakers or music [112]. However, to encode speech efficiently, the number of parameters must be reduced; and limits on parameter values are necessary. The sinusoidal speech coder is based on the knowledge that when speech is purely voiced, the frequency components of the signal correspond directly to harmonics of the pitch. As a result, the sine wave parameters correspond to the harmonic samples of the short-time discrete Fourier transform (DFT).

For this situation, Equation 9.2 simplifies to :

$$s(n) = \sum_{l=1}^{L} A_l cos(l\omega_o n + \phi_l) \qquad (9.3)$$

The amplitude estimates $A_l = |Y(l\omega_0)|$ and the phase estimates $\phi_l = \angle Y(l\omega_0)$ can be calculated from the DFT of the input speech, $Y(\omega)$. For this purely voiced case, the DFT will have peaks at multiples of ω_0, the pitch frequency. When the speech is not perfectly voiced, the DFT will still have a multitude of peaks, but at frequencies that are not necessarily harmonicly related [113]. In these cases, the sinewave frequencies are taken to be the peaks of the DFT, and the amplitudes and phases are still obtained by evaluating the DFT at the chosen frequencies. All of the above analysis is performed using a Hamming window of at least 2.5 times the average pitch period. A time window of this length is long enough to accurately resolve the individual sinusoids.

Figure 9.4 shows a block diagram of the sinusoidal transform encoder and decoder. The input speech is Fourier transformed and the peaks of the magnitude are determined by the "Sinusoidal Analysis" block. The frequencies and amplitudes are analyzed for pitch harmonics to determine pitch and voicing information. The amplitudes are transformed to the cepstral domain for more efficient coding.

The cepstral analysis does not impose an LPC model onto the spectral shape. This is claimed to result in a better fit of the spectrum in the lower frequency regions [110].

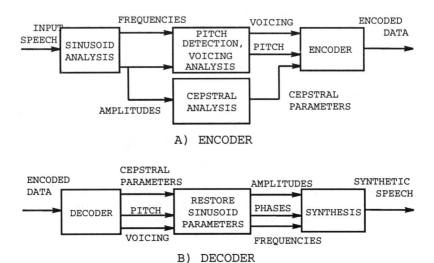

A) ENCODER

B) DECODER

FIGURE 9.4
Sinusoidal analysis and synthesis coding.

The voicing information is derived from the pitch estimator and parameterized as a cutoff frequency. Frequencies below the cutoff are voiced and harmonic. Above the cutoff, they are considered unvoiced and not harmonic, with random phase.

The decoder decodes the pitch and voicing information and inverse transforms the cepstral information. This information is combined to restore the amplitudes, frequencies, and phases for the component sine waves. The sine waves are synthesized to produce the output speech. The harmonic components (those below the voicing cutoff frequency) are synthesized to be in phase with the fundamental pitch frequency. Those components that are unvoiced (above the voicing cutoff frequency) are synthesized with a random phase.

Reference [111] contains additional information on improvements to coding efficiency and robustness to errors for the sinusoidal coder. These improvements incorporate vector quantization of subband channel energies.

9.4   Linear Prediction Vocoder

A linear prediction (LP) vocoder estimates the vocal tract using linear prediction on the speech segment (see Chapter 4). Linear prediction is also referred to as linear predictive coding (LPC). LP analysis evaluates the input speech segment and yields the impulse response of the vocal tract based on a linear model. The LPC coefficients (or a different form of the same information, such as reflection coefficients) represent the spectral envelope. In more recent coder implementations, line spectral frequencies (LSFs) have allowed a reduction in bit rate due to a more compact representation better suited to vector quantization (see Section 7.5.4).

The LP analysis does a good job of estimating, and removing, the vocal tract information in the speech signal. The remaining portion, referred to as the excitation, contains mostly the glottal pulse signal (pitch) and a noise-like, unvoiced component.

Figure 9.5 displays an FFT magnitude and LP spectrum for a voiced speech frame. The LP spectrum matches the general shape of the speech spectrum, but does not model the pitch structure. The LP spectrum models the large spectral peaks, the formants. The lower plot shows an FFT magnitude of the residual. The input speech was inverse filtered by the LP filter to obtain the residual. The resulting excitation spectrum is much whiter (flatter) than the original spectrum and retains the strong periodic nature of the original speech signal as is evident by the pitch harmonics.

Figure 9.6 displays an FFT magnitude and LP spectrum for an unvoiced speech frame. The lower graph shows the FFT magnitude of the residual signal. The inverse LP filtering has removed the general spectral shape, and the spectrum of the residual is flat.

The major problem in using the full excitation signal in practice is the large number of bits required to transmit it [13]. A great deal of effort has been directed at reducing the coding requirements of the excitation signal. The LP-based analysis-by-synthesis coders of Chapter 10 are geared toward accurate, efficient representation of the time-domain residual signal.

The residual excited linear prediction (RELP) coder uses LP analysis to remove the spectral envelope and codes only the baseband, or low frequency, portion of the residual. One particular approach [131] low-pass filters the residual to 0 to 1000 Hz and encodes the magnitudes and

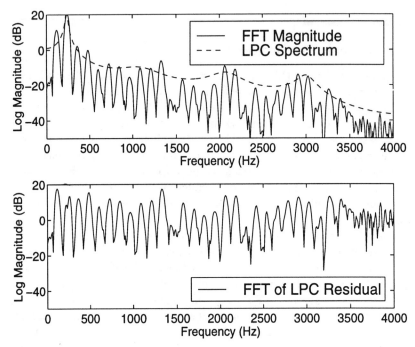

FIGURE 9.5
LP spectrum and residual spectrum for voiced speech frame.

phases of the FFT of the residual. At the decoder, this base-band residual is replicated at higher frequency bands to generate a full bandwidth residual. This synthesized residual is then passed through the LPC filter to synthesize the speech.

The LPC coder based on the classical two-state voicing decision results in the lowest bit rate for the residual excitation. The conceptual basis behind this method is the simple production model of Figure 2.14. For each frame, the speech is classified as either voiced or unvoiced and, as such, requires only one bit for coding the voicing.

Figure 9.7 displays a block diagram for the LPC encoder. The input speech is sampled and segmented into frames. For each frame, an LP analysis is performed to represent the spectral envelope. Typical LP orders range from 10 to 14 depending on the bandwidth of the speech and bit-rate limits. The pitch period is estimated by a separate algorithm. The voiced/unvoiced decision can utilize information about the

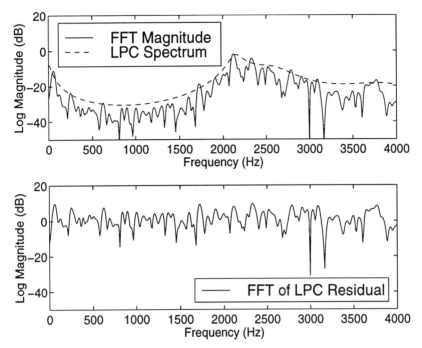

FIGURE 9.6
LP spectrum and residual spectrum for unvoiced speech frame.

harmonic structure (or lack of) in the spectrum from the pitch estimation, or the decision can be based on simple parameters such as the energy of the frame and number of zero crossings of the time waveform in the frame. Voiced speech tends to have higher energy and a lower number of zero crossings than unvoiced speech.

The LPC decoder of Figure 9.8 shows an implementation of the source-filter speech production model. The pulse generator produces a periodic waveform at intervals of the pitch period. The noise generator outputs a random sequence of white (equal power across the spectrum) noise. The voicing information controls the switch that decides whether to select the periodic (voiced) or random (unvoiced) excitation signal. The excitation signal is then frequency shaped by the LPC filter and multiplied by the gain to produce the correct energy (signal amplitude) of the output synthesized speech.

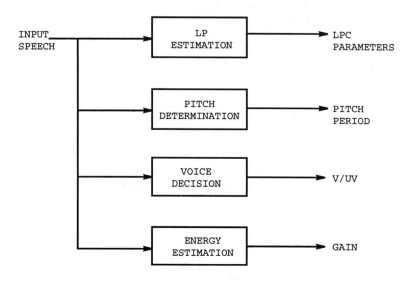

FIGURE 9.7
Linear predictive coding (LPC) encoder.

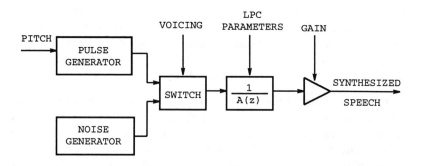

FIGURE 9.8
LPC decoder.

9.4.1 Federal Standard 1015, LPC-10e at 2.4 kbit/s

The U.S. Department of Defense adopted a 2.4 kbit/s LPC vocoder in 1982, and with later modifications, it became the Federal Standard 1015, LPC10e [15, 182]. For the required 8 kHz sample rate, the frame length is 22.5 ms, and 54 bits/frame achieve the 2.4 kbit/s total rate.

To estimate the pitch, the average magnitude difference function (AMDF) is computed as:

$$\text{AMDF}(m) = \sum_{n=0}^{N-1} |s(n) - s(n+m)| \tag{9.4}$$

for the speech samples, $s(n)$, over the window of length N. For periodic speech, the AMDF will have a valley for the values of m near the pitch period. The range of allowable pitch frequencies is limited to 50 to 400 Hz. The AMDF is similar to the autocorrelation method of pitch estimation but is simpler computationally, using differences and absolute magnitudes instead of multiplications. Six bits are used to code the pitch value.

For voicing, two decisions are made for each frame, one at the beginning and another at the end of the frame. The decision takes into account the low frequency energy, the zero-crossing count, the first two reflection coefficients, and the ratio of the AMDF maximum to minimum. The first reflection coefficient provides a measure of spectral tilt. For voiced speech, the lower frequency components are of greater magnitude than the higher, resulting in a significant spectral tilt. The second reflection coefficient is computed from lowpass (800 Hz) filtered speech and is a measure of spectral peakedness. A strong peak character in the low frequency spectrum is an indicator of voicing. The voicing decision is performed by a linear discriminant classifier. The classifier adapts to different input noise levels by estimating the input SNR and using different coefficients in the linear combination of terms. Reference [15] details the voicing classifier.

Both the pitch estimates and the voicing decisions are adjusted with dynamic programming. For the pitch values, the dynamic programming tracking removes occurrences of pitch halving or doubling errors. The dynamic programming reduces spurious switching between voicing decisions.

The LPC analysis is tenth order and applies the covariance method to estimate the parameters. A Cholesky decomposition (see Section 4.2.2) solves the system of equations. The LPC coefficients are represented as reflection coefficients for indices 3 through 10 and as log-area ratios for the first two coefficients. The full tenth-order list of parameters is coded for voiced segments, but only the first 4 (4th order LPC representation) are coded for unvoiced speech. The first 4 coefficients are coded with 5 bits per coefficient, while coefficients 5 through 8 are coded with 4 bits (for voiced speech only). The ninth coefficient uses 3 bits, and the tenth, 2 bits. For unvoiced segments where only the first 4 are coded, the extra bits are used for channel error coding.

For intelligibility tests, the LPC-10e coder scored a 90% on the diagnostic rhyme test (DRT). Diagnostic acceptability measure (DAM) testing of the LPC-10e resulted in a score of 48. In a later comparison to more recent coders, the LPC-10e scored 2.2 as a mean opinion score (MOS) relative to a 3.1 for the Federal Standard 1016 CELP and a 3.3 for the Federal Standard MELP [95].

The LPC-10e algorithm, while low in complexity, produces coded speech with a somewhat buzzy, mechanical quality. At the 2.4 kbit/s rate, it has been recently replaced by the MELP Federal Standard. Appendix A includes an on-line location that contains examples of speech coded and decoded by LPC-10e and downloadable source code of a software implementation in either Fortran or C.

Chapter 10

Linear Prediction Analysis by Synthesis

This chapter presents several linear prediction (LP) coding methods that incorporate analysis-by-synthesis (AbS). These coders model the vocal tract (short-term spectrum) with linear prediction. In the encoder, the optimal parameterization of the excitation (LP residual) is determined using analysis by synthesis. The optimal excitation parameters are those that produce synthesized speech that most closely matches the original speech.

Analysis by synthesis is a powerful method used in parameter estimation. It is a closed-loop approach that incorporates a version of the decoder in the encoder. Parameters are determined, or adjusted, by minimizing an error signal. The error signal is the difference between the original speech and the locally generated synthetic version. The scheme is closed-loop because of the iterative nature of the refinement of the parameters to be estimated. The updated parameters are used to synthesize the output; the error between the synthesized output and the original guides the next round of parameter refinement.

Multi-pulse (MP) coders specify the excitation signal through the location and amplitude of nonuniformly spaced pulses. Regular pulse excitation (RPE) coders use pulses to approximate the excitation signal, but restrict the pulse locations to even spacing. The excitation signal is determined by the pulse spacing interval, the location of the first pulse, and the pulse amplitudes.

Code excited linear prediction (CELP) uses a codeword to specify a vector that is the time-domain excitation signal. The coding of the excitation is a form of vector quantization. With recent advances in coding quality, CELP implementations dominate speech coding at bit rates from 5 to 12 kbits/s.

10.1   Analysis by Synthesis Estimation of Excitation

Figure 10.1 displays a block diagram for a general LPC-based analysis by synthesis encoder (MP, RPE, CELP). The "Improve Parameters" block outputs new excitation parameters based on the reduction of a perceptually weighted error. An initial set of parameters starts the process, and the parameters are adjusted to reduce the error signal. The "Generate Excitation" and "LPC Filter" blocks form the local decoder, and together produce the synthetic signal for comparison.

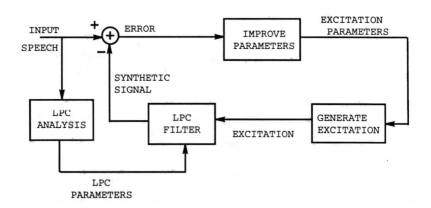

FIGURE 10.1
Generalized Analysis by Synthesis encoder.

In practice, the error is multiplied by a frequency weight $W(z)$ as:

$$E(z) = W(z)(S(z) - S_{synth}(z)) \qquad (10.1)$$

where $W(z)$ depends upon the LP frequency response as:

$$W(z) = \frac{1 - A(z)}{1 - A(z/\gamma)} \qquad (10.2)$$

and $A(z)$ is the LPC polynomial so that:

$$W(z) = \frac{1 - \sum_{i=1}^{p} a_i z^{-i}}{1 - \sum_{i=1}^{p} \gamma^i a_i z^{-i}} \qquad (10.3)$$

The constant γ is chosen to be near 0.8, based on empirical results. The shape of the weighting filter is similar to the inverse of the input speech spectra. The purpose of the weighting is to reduce the contribution of the spectral peaks (associated with the formants) in the error calculation. Otherwise, the large magnitudes of the peaks would dominate the error and, in turn, the parameter fitting, at the expense of the frequency regions between the peaks.

10.2   Multi-Pulse Linear Prediction Coder

Multi-pulse (MP) coders [4, 147, 5, 148] model the vocal tract-induced, short-term spectrum with an LP analysis, and then approximate the residual as a sequence of pulses that are not evenly spaced. This approach avoids the sometimes difficult or inaccurate decision on whether the frame is completely voiced or unvoiced. The location and amplitude of the pulses is determined using analysis by synthesis.

For MP, the pulse locations and amplitudes are determined sequentially, beginning with the first pulse position and amplitude that most closely approximates the residual function. Given the first, the second is positioned so as to further reduce the error. This sequential approach, while computationally attractive, is not optimal. To improve the fit, some implementations reestimate the pulse amplitudes while holding the pulse positions constant. Most implementations use 10 to 12 pulses to approximate a 10 ms segment of the residual [96]. The parameters for the residual are updated every 5 to 10 ms, more frequently than the LP parameters for which a typical update rate would be 15 to 20 ms. Reference [96] gives the derivation of the equations for computing the pulse positions and amplitudes, the reestimation of the optimal amplitudes after the positions are set, and methods of coding the positions and amplitudes.

Reference [147] proposed introducing the long term predictor (LTP) as shown in Figure 10.2. The LTP is typically implemented as a single lag filter (although 3-tap versions are common). The single-tap version has only two parameters: the gain and the delay. The LTP serves to repeat the excitation sequence. Without the LTP, most pulses are used to model the glottal pulses across the frame for voiced speech. With the LTP, for voiced speech, the lag, or delay, of the LTP corresponds to

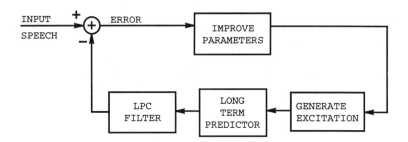

FIGURE 10.2
Analysis-by-synthesis linear prediction coder with addition of
long term predictor (LTP).

the pitch period and allows a repeating of the pulses associated with the
glottal pulse. The remaining pulses are available to model the rest of the
structure in the excitation. The lag of the LTP can be determined by
finding the maximum of the autocorrelation of the residual signal, in the
same manner as the autocorrelation method of pitch estimation. The
LTP parameters are set once, outside the normal closed-loop iterations
used to determine the pulse locations and amplitudes. In other coders
discussed below, the LTP parameters are often optimized inside the
closed-loop excitation parameter estimation.

10.3   Regular Pulse Excited LP Coder

Regular pulse excited (RPE) LP coding [97] is a form of multi-pulse
with additional constraints placed on the positions of the pulses. The
pulse positions are evenly spaced, so determining the location of the first
pulse sets the location of all the pulses. Practical implementations use
10 to 12 evenly spaced pulses over a 5 ms segment of the excitation.

The pulse positions and amplitudes are determined in a closed-loop
analysis by synthesis scheme. The optimal pulse amplitudes that minimize the error can be computed efficiently [97]. The spacing between
pulses is only 3 to 4 samples for an 8 kHz sampling rate, so the optimal
amplitudes are computed for each possible initial pulse position and the
one resulting in the overall minimum error is selected. As with the MP

coder, the addition of a long term predictor improves the performance of the system with only a slight increase in complexity.

10.3.1   ETSI GSM Full Rate RPE-LTP

The Groupe Speciale Mobile (GSM) of the European Telecommunications Standards Institute (ETSI) standardized an RPE coder for mobile cellular applications [180]. The standard uses 13 kbit/s for speech coding and the remainder of the 22.8 kbit/s channel for error control.

Figure 10.3 displays simplified block diagrams for the encoder and decoder. In the LTP estimation loop and the RPE grid selection loop, all the parameters are coded and decoded (as is necessary for analysis by synthesis) before being used for reestimation. This is necessary for the small degradations and errors associated with quantization to be accurately represented in the encoder's version of the decoder.

An eighth order LP analysis is performed every 20 ms. The LP coefficients are coded as log area ratios (LARs) with 36 bits, more bits being assigned to the initial coefficients. The LP parameters are used to inverse filter the input speech to obtain the residual.

The pulse amplitudes and grid position for the excitation are estimated every subframe of 5 ms. The amplitudes are normalized by the maximum of the segment. The maximum is coded logarithmically with 6 bits, and the samples are coded uniformly with 3 bits each.

The residual signal is filtered by the LTP. The delay of the LTP is coded with 7 bits, while the gain uses 2 bits.

Subjective listening tests resulted in a mean opinion score (MOS) of 3.47. The GSM Full Rate RPE-LTP has been superseded by the Enhanced Full Rate (EFR) CELP coder that offers improved speech quality at a lower bit rate. Documentation for the standard [180] is available on-line at the site listed in Appendix A.

10.4   Code Excited Linear Prediction Coder

In linear prediction coding, filtering each speech segment with the inverse LP filter yields a residual signal. If the residual signal were used as the excitation of the filter, the output would be identical to the original windowed speech segment. If the excitation closely resembles

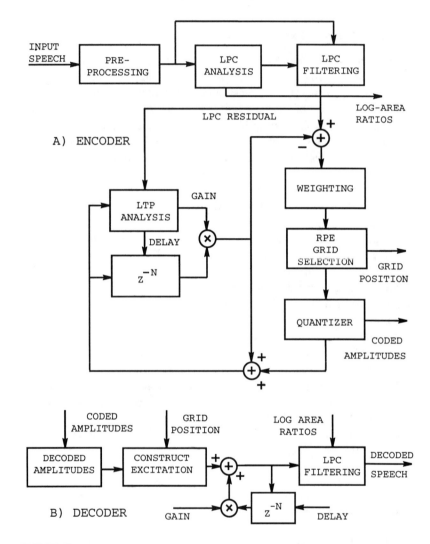

FIGURE 10.3
GSM Full-Rate Regular Pulse Excited coding standard.

the residual, then the resulting reconstructed speech signal yields high speech quality.

The innovation of the CELP coder is having prearranged excitations available and using the excitation that best matches the residual signal [142, 98, 5, 1]. Each of the excitations has a codeword associated with it, and only the codeword (and possibly a LTP delay and gain) need to be transmitted for reconstruction of the residual.

10.4.1 CELP Concept

CELP is an analysis by synthesis method of encoding. The CELP approach is most easily conceptualized with a block diagram of the original formulation [142]. In Figure 10.4, codebook sequence $c_k(n)$ is filtered through the synthesis path of the local version of the decoder incorporated into the encoder. The synthesis path includes the gain to properly scale the codebook sequence, the LTP filter $1/B(z)$, and the short term predictor (STP) filter $1/A(z)$.

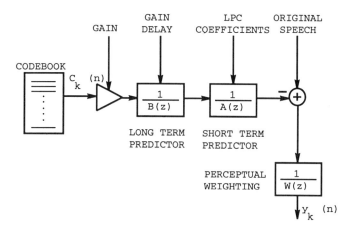

FIGURE 10.4
Code excited linear prediction (CELP) scheme, minimize $y_k(n)$ by selecting best codebook entry.

The LTP predictor is characterized by the delay and one or three coefficients corresponding to a one- or three-tap filter as shown by the equation:

$$\frac{1}{B(z)} = \frac{1}{1 - \sum_{i=-I}^{I} b_i z^{-(\tau+i)}} \tag{10.4}$$

where I is the order of the predictor (zero for a one-tap, one for a three-tap); the b_i terms are the filter coefficients; and τ is the delay. In the closed-loop estimation of the LTP gain and delay, the LTP can be viewed as an adaptive codebook [91]. In this view, the adaptive codebook is populated by time-shifted versions of the past excitation, and the delay, τ, becomes the index to the appropriate codebook entry. The search locates the best fit of the recent synthetic excitation to the excitation of the current subframe.

The short term predictor is characterized by the LPC coefficients. The synthesized version is subtracted from the original speech and the difference is weighted by the perceptual filter, $W(z)$. The structure and function of the three filters – LTP, LPC, and perceptual weighting – is the same as the discussion for multi-pulse in Section 10.2.

The entire process just described is repeated for each index, k, of the codebook sequence, $c_k(n)$, to determine the sequence, $y_k(n)$, with the minimum total energy. The sequence with the minimum total energy provides the best-match excitation, and that index is stored and transmitted to the decoder. In the original formulations, the codebook is populated by random sequences with a Gaussian distribution and unit variance. The decoder synthesizes the speech in exactly the same manner as had been done at the encoder.

10.4.2   CELP Computational Efficiency Improvements

The primary barrier to practical CELP implementations was the large number of computations required to filter each codebook entry by the three filters. Rearranging the equations and utilizing highly structured, computationally efficient codebooks has reduced the burden to manageable levels.

A simple rearrangement, used in all coders, is to move the weighting filter, $W(z) = A(z)/A(z/\gamma)$, to both branches that lead into the summation. The result is shown in Figure 10.5. This reduces the complexity by requiring the input speech to be perceptually weighted (only one computation) but removes that computation from the bottom branch and, consequently, for each codeword of the codebook.

Reference [91] describes a number of computation reductions and fast algorithms, the most significant are mentioned here. Sparse codebooks

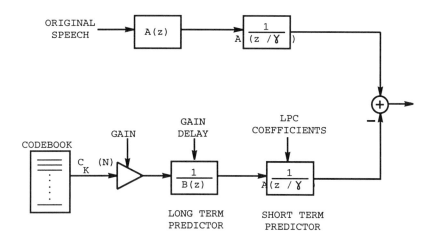

FIGURE 10.5
Reorganized CELP processing flow to reduce computation.

are examples where the codewords have a significant

percentage of zero values (as much as 90 to 95%). Surprisingly, performance is nearly the same relative to a full stochastic codebook. The computation savings is significant by the reduction of the number of multiplications for the computation of inner products.

Binary (-1, 1) and ternary (-1, 0, 1) valued codebooks have been shown to provide good performance. Convolution of these codewords with CELP filters offers much lower complexity because the multiplications of inner products are reduced to additions and subtractions.

Overlapping codebooks offer several implementation advantages. Overlapping codebooks are not composed of independent vectors, but have all the values stored in one array. The first codeword is the first segment of the array, and the second codeword begins somewhere in the first word and extends past the first, farther into the array. The structure is shown by Figure 10.6, where N is the total number of codewords, "Length" is the number of elements in one codeword, and "Shift" is the offset between adjacent codewords.

Overlapping codebooks require less storage space. Because adjacent codewords have mostly the same values, efficient algorithms can exploit this dependency to reduce the required computations. Kleijn and colleagues reported in [91] that as codebook size increases, small shift sizes

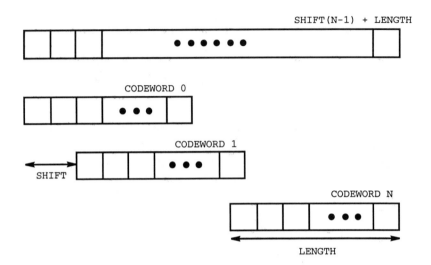

FIGURE 10.6
Structure of overlapping codebook and extraction of individual codewords.

can nearly achieve the performance of a completely independent codebook. A shift of two was cited as appropriate.

10.4.3   Adaptive Postfiltering

Adaptive postfiltering is a component of most CELP systems, as well as most other current LPC-based approaches to low rate coding. The most commonly used scheme is credited to Chen in [19].

The purpose of the postfilter is to lower the perceived noise in the synthesized signal by attenuating the signal in the spectral valleys. To accomplish this, it uses a variant of the frame-by-frame LPC information to adapt to the changing spectral shape. If only the variant of the LPC filter were used to boost the formants and lower the valleys between formants, the resulting spectrum would have increased spectral slope and sound as if it had been lowpass filtered. To compensate for the spectral tilting, a sort of deemphasized inverse LPC filter is used, in which the parameters have been modified.

The usual form for the adaptive postfilter is:

$$H(Z) = \frac{(1 - \mu z^{-1})(1 - \sum_{i=1}^{p} a_i \beta^i z^{-i})}{1 - \sum_{i=1}^{p} a_i \alpha^i z^{-i}} \qquad (10.5)$$

where p is the LPC order and a_i terms are the LP filter coefficients. The μ, β, and α quantities are set to perform the desired level of postfiltering. The denominator is the LPC-like filter to boost the formants, the numerator is the deemphasized LP inverse to compensate for most of the spectral tilting, and the $1 - \mu z^{-1}$ term is a highpass component to compensate for the rest of the spectral tilt. Typical values for the postfilter factors are $\mu \approx 0.3$, $\beta \approx 0.6$, and $\alpha \approx 0.9$. The relation between α and β controls the amount of formant enhancement.

10.4.4 Federal Standard 1016, CELP at 4.8 kbits/sec

The U.S. Department of Defense Federal Standard 1016 [183, 16] was adopted in 1991. The CELP algorithm operates at a frame rate of 30 ms with 4 subframes of 7.5 ms. The excitation is reestimated each subframe. The LTP, or adaptive codebook, contains 256 codewords and is searched over delays from 20 to 147 samples. The fixed codebook of sparse, ternary values contains 512 codewords of 60 samples in length. The STP incorporates a tenth-order LP predictor. The analysis window for the LP is a 30 ms Hamming window. The total throughput delay is 105 ms for the CELP algorithm.

In the testing of standard coders reported in [95], the CELP standard registered scores of 65 for the diagnostic acceptability measure (DAM), 3.1 for the MOS, and 91% for the diagnostic rhyme test (DRT). These test numbers were scored under quiet conditions. Reference [95] compares the CELP against three other standards (MELP, CVSD, and LPC-10e) over a wide range of noise conditions and other testing paradigms.

Appendix A includes an on-line location that contains examples of speech coded and decoded by the FS1016 CELP algorithm, and downloadable source code in Fortran or C for a software implementation.

10.4.5 ITU-T G.728 Low Delay CELP at 16 kbit/s

When speech coding algorithms are used for communication systems, the coding delay can become a problem. The coding delay is the time it takes for a time sample in the speech to be processed through both the encoder and decoder. This accounting scheme ignores the additional delays due to transmission considerations. The coding delay results from data buffering at the encoder and computational delays. The delay can be several frames in duration, often from 50 ms to over 100 ms.

The G.728 CELP [189, 20] standard approaches the coding problem in a significantly different manner to reduce its coding delay to less than 2 ms. Typical CELP coders buffer a frame of speech at the encoder, perform LP analysis, and transmit the LPC and excitation information to the decoder. This process is referred to as forward adaptive. In low delay CELP (LD-CELP), only the excitation is transmitted. The coefficients of the STP are updated at the decoder based on prior decoded speech samples by backward adaptive prediction.

Figure 10.7 displays the block diagram of the LD-CELP algorithm. The LD-CELP has no long term predictor, but uses a high order, 50, short term predictor. The high order STP can model the pitch structure in the excitation in place of an LTP.

The quantized data are windowed with an asymmetric Barnwell window [7] for the LPC computation. This method allows an efficient recursive computation of the LPC parameters. With the asymmetric window, the most recent speech samples are weighted more heavily in the LP analysis.

The frame is 2.5 ms long (20 samples) with 4 subframes. That yields an excitation vector only 5 samples long. The gain factor for the excitation, the output of the box labeled as "Backward Gain Adaptor," is predicted with a tenth-order linear predictor. The gain is predicted in the log domain by the backward adaptive scheme.

The codebook uses a 10-bit, shape-gain vector quantization approach, with 7 bits for the shape and 3 for the gain. The codebook vectors are trained, instead of the usual random distribution for fixed codebooks in CELP algorithms. The perceptual weighting filter, as in the general CELP discussion, includes tenth-order LP parameters. An adaptive postfilter renders the quantization noise less obvious.

Reported MOS scores are only slightly less than 4, with a score of 4 considered to be toll-call communications quality.

10.4.6   ITU G.723.1 Algebraic CELP/Multi-Pulse Coder at 5.3/6.3 kbit/s

The standard ITU G.723.1 [187] coder is designed for video conferencing and voice-over-Internet applications. It is specified as part of the

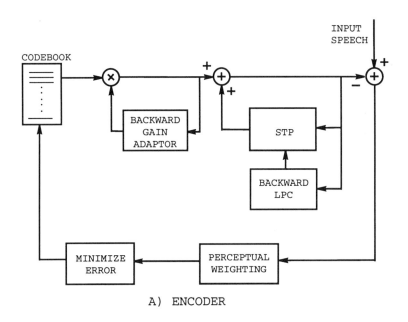

A) ENCODER

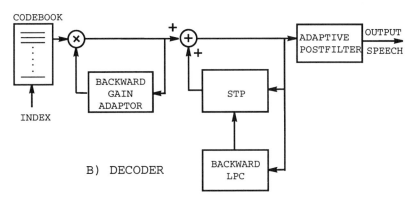

B) DECODER

FIGURE 10.7
ITU G.728 standard low delay CELP coder.

audio coding component of ITU recommendations H.323 and H.324 on video telephony.

The coder is a dual rate, switchable implementation. The LP information is coded by LSFs. The excitation is coded by CELP or MP methods. The Algebraic CELP (ACELP) encodes speech at 5.3 kbit/s. (A slightly different algebraic codebook structure is explained in the next section.) The 6.3 kbit/s multi-pulse maximum likelihood quantization (MP-MLQ) offers slightly better quality at the higher bit rate. Both coders are included in the coder and decoder. Switching between coders can occur at any frame boundary. The frame length is 30 ms with 4 subframes. The total algorithmic delay is 37.5 ms.

10.4.7 ETSI GSM Enhanced Full Rate Algebraic CELP at 12.2 kbit/s

The GSM Enhanced Full Rate (EFR) Algebraic CELP (ACELP) [179, 77] encodes speech at a 12.2 kbit/s rate. The EFR was designed for mobile digital cellular communications. As such, 10.6 kbit/s are utilized for error control channel coding for a total bit rate of 22.8 kbit/s.

The GSM EFR operates as a general CELP coder as described in Section 10.4. The algorithm divides the 20 ms frame into four 5 ms subframes. A tenth-order LPC analysis, including an asymmetric 30 ms window, is performed twice per frame. The first window has most weight concentrated on the second subframe, and the second window places the most significance on the fourth subframe. The LP parameters are converted to line spectral pairs (LSPs). First-order moving average (MA) prediction is applied to the LSPs, and the prediction residual is quantized with split matrix quantization. The two sets of LSPs are quantized with a total of 38 bits.

The algorithm incorporates an initial open-loop LTP lag search, followed by closed-loop, subinteger refinement. The open-loop search is done twice per frame, and the closed-loop search is repeated for each subframe. The lag is quantized with 9 bits for the first and third subframe, and differentially with 6 bits for the second and fourth.

The algebraic codebook is structured around 5 *tracks*. The tracks determine the allowable positions for the 10 nonzero pulses in each subframe of 40 samples. Each pulse amplitude can be +1 or −1, and each track has 2 pulses. Both pulses can be positioned in the same location to produce a pulse amplitude of +2 or −2. Table 10.1 lists the allowable pulse positions for each of the five tracks. In each track, the allowable

Track	Pulses	Allowable Positions
1	p0, p1	0, 5, 10, 15, 20, 25, 30, 35
2	p2, p3	1, 6, 11, 16, 21, 26, 31, 36
3	p4, p5	2, 7, 12, 17, 22, 27, 32, 37
4	p6, p7	3, 8, 13, 18, 23, 28, 33, 38
5	p8, p9	4, 9, 14, 19, 24, 29, 34, 39

Table 10.1 Allowable pulse positions for GSM Enhanced Full Rate

positions are spaced on a grid of 5 samples.

The pulse positions are optimized by a nonexhaustive analysis by synthesis search to minimize the perceptually weighted error. First, the globally best position for p0 is determined. Then, five iterations determine the position of pulse p1, with each iteration incorporating a series of nested loops to search for the positions of the other four pulse pairs. A total of 35 bits is required to code the fixed codebook information.

At the decoder, the synthesized speech is filtered with an adaptive postfilter as described in Section 10.4.3. Decoded speech quality is high. Subjective MOS scores are near or above 4. The quality has been reported as at or above that of the standard wireline 32 kbit/s ADPCM [77].

Detailed information concerning the GSM EFR, including the standard, is available online at the site listed in Appendix A.

10.4.8 IS-641 EFR 7.4 kbit/s Algebraic CELP for IS-136 North American Digital Cellular

The Interim Standard 641 (IS-641) ACELP [192, 73] is a coding standard designed for the North American digital cellular IS-136 Time Division Multiple Access (TDMA) system. The IS-641 standard is very similar to the GSM EFR coder, both having been developed by the same group of researchers in the same time period. The IS-641 includes 5.6 kbit/s of channel coding for total rate of 13.0 kbit/s.

The LPC analysis is performed only once per 20 ms frame. The LSP is split vector quantized with 26 bits per frame. The adaptive codebook (LTP) lag is coded with 8 bits on the first and third subframes, and with 5 bits for the differential for the second and fourth subframes.

The algebraic codebook is structured in the same manner as the GSM EFR; however, only 4 pulses per subframe are allowed. The allowable positions are the same as shown in Table 10.1. The codebook information is coded with 17 bits for each subframe.

The IS-641 EFR coder reportedly offers toll-quality or near toll-quality in error-free conditions [73].

10.4.9   ETSI GSM Adaptive Multi-Rate Algebraic CELP from 4.75 to 12.2 kbit/s

The European Telecommunications Standards Institute (ETSI) has standardized an adaptive multi-rate (AMR) coder for use in digital cellular applications [178]. In cellular applications, obtaining the best quality speech at the receiving end is a trade-off between allocating bits to the speech coder or to the error channel coding. The split of the allocation depends on how error free or error prone the channel is. The AMR system adapts to the channel conditions by selecting the appropriate mode (full or half rate) and the speech coding rate that allows sufficient error protection for the error level of the channel.

The GSM AMR encodes the speech by Algebraic CELP (ACELP) at one of eight possible bit rates: 4.75, 5.15, 5.90, 6.7, 7.4, 7.95, 10.2, or 12.2 kbit/s. The high rate coder, 12.2 kbit/s, defaults to the GSM Enhanced Full Rate (EFR) coder. The coder will operate in both the full rate (22.8 kbit/s) and half rate (11.4 kbit/s) channels. One of eight rates (as listed) is used to code the speech information, the remaining channel bandwidth is used for error protection (channel coding). For a noisy channel, prone to errors, a low rate speech coder is selected, leaving significant bandwidth for channel error coding. For clean channels, a high rate coder is chosen that provides high quality decoded speech.

As mentioned, the 12.2 kbit/s coder is the GSM EFR coder. For the other bit rates, a tenth order LP analysis is performed once per 20 ms frame with an asymmetric window (more heavily weighted with most recent time samples). The equations of the autocorrelation method are solved by the Levinson-Durbin recursion. The LP information is quantized in the LSP representation. The LSP vector is predicted with a first-order predictor. The residual is quantized with split VQ (SVQ). The 10 LSPs are split into vectors of length 3, 3, and 4. Each subvector is coded with 7 to 9 bits, depending on the overall bit rate for speech coding.

The 20 ms frame is divided into 4 subframes. The LP residual is calculated for each subframe. The adaptive codebook (LTP) gain and lag are determined for each subframe.

Mode	LSP	Pitch Lag	Fixed Codebook	Gains	Total
10.2 kbit/s	26	8, 5, 8, 5	31, 31, 31, 31	7, 7, 7, 7	204
7.95 kbit/s	27	8, 6, 8, 6	17, 17, 17, 17	9, 9, 9, 9	159
7.40 kbit/s	26	8, 5, 8, 5	17, 17, 17, 17	7, 7, 7, 7	148
6.70 kbit/s	26	8, 4, 8, 4	14, 14, 14, 14	7, 7, 7, 7	134
5.90 kbit/s	26	8, 4, 8, 4	11, 11, 11, 11	6, 6, 6, 6	118
5.15 kbit/s	23	8, 4, 4, 4	9, 9, 9, 9	6, 6, 6, 6	103
4.75 kbit/s	23	8, 4, 4, 4	9, 9, 9, 9	8, 0, 8, 0	95

Table 10.2 Bit allocation by frame for GSM AMR coder [178]. Comma-separated values in a table entry denote bit allocation for each subframe.

The fixed, algebraic codebook is similar to that of the GSM EFR. However, each lower bit rate coder uses a codebook with fewer tracks to reduce the number of bits required to encode the index. The lower rates include only one pulse per track.

The bit allocations for the different bit rates are displayed in Table 10.2 [178]. The comma separated values are the bit allocations for the subframes, four per frame.

At the decoder, the LSPs are interpolated for each subframe. The adaptive and fixed codebook contributions are weighted by their gains. Standard adaptive postfiltering is applied to enhance the speech.

Chapter 11

Mixed Excitation Coding

Code excited linear prediction (CELP) coding methods perform well at bit rates near 5 kbit/s and above. However, as system requirements lower the bit rate below that level, the quality of CELP output speech declines precipitously. This is primarily due to the fact that not enough bits are available to represent a sufficient number of codes for the excitation. Or viewed differently, the coder expends too many bits attempting to copy perceptually unimportant time-domain details of the excitation signal.

Several coding approaches, including Multi-Band Excitation (MBE), Mixed Excitation LP (MELP), Harmonic Vector Excitation (HVXC), and Waveform Interpolation (WI), have evolved to provide good speech quality at bit rates from 4 kbit/s down to 1.2 kbit/s. They all incorporate innovations to efficiently model the excitation. To improve speech quality over the classic two-state, voiced/unvoiced LP coder (see Section 9.4), these coders include both harmonic and noise-like components simultaneously in the modeling and regeneration of the excitation signal. The presence of both components results in the label of *mixed excitation*. These coders represent the current state of the art in low rate coding. As such, these coders and their variants are topics of intense current research efforts.

11.1 Multi-Band Excitation Vocoder

The Multi-Band Excitation (MBE) vocoder [63, 64, 62, 65] is a frequency-domain coder that incorporates an innovation to better model the excitation. Because many speech segments are not purely voiced or

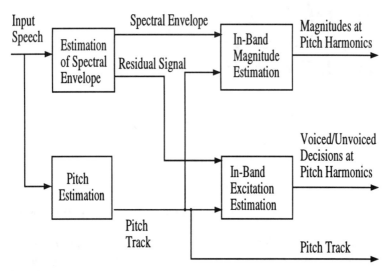

FIGURE 11.1
Speech analysis in Multi-Band Excitation (MBE) encoder.

unvoiced, a single voiced/unvoiced (V/UV) decision per frame of the classic two-state vocoder is not completely accurate in those cases. The MBE excitation is mixed, allowing both harmonic and random components in a single frame of speech. For voiced speech, a periodic sequence of excitation impulses corresponds to a periodic sequence of impulses in the frequency domain, spaced at the harmonics of the pitch. The MBE model divides the spectrum into subbands at multiples of the pitch frequency. The manner in which the MBE vocoder represents the vocal tract frequency information can be thought of as a channel vocoder that has all channels centered at harmonics of the pitch frequency. The MBE model allows a separate V/UV decision for each frequency channel (or group of channels) in each frame of speech. This allows a more faithful representation of the excitation signal than with single V/UV decision vocoders.

11.1.1   Multi-Band Excitation Analysis

Figure 11.1 shows the analysis portion of the MBE vocoder. An accurate estimation of the pitch frequency track is essential to properly position the higher pitch harmonics for accurate speech synthesis. The pitch estimate for each frame is determined by the following three part process:

1. Using an algorithmically efficient autocorrelation method, a rough pitch period versus error calculation is made for the integer sample values within an allowable range of pitch periods.

2. Dynamic programming is used to smooth the pitch frequency track from the error calculations with the restriction that the pitch track must change "slowly" on a frame-by-frame basis.

3. The pitch period is then refined by a frequency-domain matching method to obtain a more precise pitch estimate (noninteger sample multiples).

The autocorrelation function used for the initial estimate is normalized by the windowed speech as in the following:

$$e(p) = \frac{\left[\sum_{m=-N/2}^{N/2} s^2(m)w^2(m)\right] - p\sum_{n=-N/2p}^{N/2p} R(np)}{\left[\sum_{m=-N/2}^{N/2} s^2(m)w^2(m)\right]\left[1 - p\sum_{m=-N/2}^{N/2} w^4(m)\right]} \quad (11.1)$$

where the analysis frame length is $N+1$, and $w(m)$ is the speech window, normalized such that the summed energy of the window is unity as follows:

$$\sum_{n=-N/2}^{N/2} w^2(n) = 1 \quad (11.2)$$

The autocorrelation, $R(n)$ is:

$$R(n) = \sum_{m=-N/2}^{N/2} s(m)w^2(m)s(m+n)w^2(m+n) \quad (11.3)$$

The dynamic programming pitch tracker effectively removes occurrences of pitch halving and doubling. If the actual pitch period is P for a segment of speech, Equation 11.1 will produce a local (and possibly global) maximum at $2P$. Essentially, this case matches two pitch periods of the original speech waveform to two subsequent pitch periods, separated by a shift of two pitch periods. The tracker is biased towards lower pitch period estimates and pitch tracks that change slowly. It eliminates the doubling because the two values of $e(p)$ will be similar at P and $2P$.

The subbands are chosen to be centered at harmonics of the pitch frequency with one subband per harmonic. Each subband or, more commonly, a group of subbands is then classified as either voiced or unvoiced.

This is done by computing the spectral error in fitting the magnitude of the original speech spectrum to that of a purely harmonic signal over the width of the group of subbands. In the purely harmonic signal, the harmonics are spaced at pitch frequency. The error is computed as follows:

$$error_j = \frac{\sum_{k_{first}}^{k_{last}} |S(n) - S_{synth}(n)|^2}{\sum_{k_{first}}^{k_{last}} |S(n)|^2} \qquad (11.4)$$

where $S(n)$ is the original speech spectrum, and k_{first} and k_{last} are the first and last harmonics in the jth band. $S_{synth}(n)$ is the reconstructed speech assuming a voiced harmonic at each pitch spacing.

If the match is good, the error will be low, and the frequency bin will be considered voiced. If the match is poor, a high error level will be detected for that bin, and it will be marked unvoiced. The low or high error threshold was experimentally determined.

The spectral magnitude estimates for the voiced bins are computed by summing the values of the original spectrum over the frequency range of the bin, normalized by the square of the magnitude of the window. For unvoiced bins, the magnitude is calculated as the root mean square (RMS) value of the spectrum over the frequency bin.

The analysis model parameters for each frame include the pitch period, the voiced/unvoiced decisions for each bin (or groups of bins), the spectral envelope magnitudes, and phase estimates for the voiced bins. Phases are not required for the unvoiced bins. In some lower bit-rate implementations, the phases are not transmitted, but are synthesized at the decoder.

Figure 11.2 displays the spectral magnitudes and voicing decisions for each harmonic for a frame of speech. For reference, the magnitude of the DFT of the segment is also plotted. The speech frame is mixed excitation from the phoneme /zh/, as in the center consonant sound in the word "vision." The low frequency group of harmonics, up to about 1000 Hz, is clearly voiced. The harmonic peaks of the discrete Fourier transform (DFT) are well formed and evenly spaced, and have been classified as voiced by the MBE analysis algorithm. The rest of the spectrum displays mixed excitation characteristics. For the most part, the DFT magnitude and the MBE magnitudes appear as unvoiced in the high frequency range.

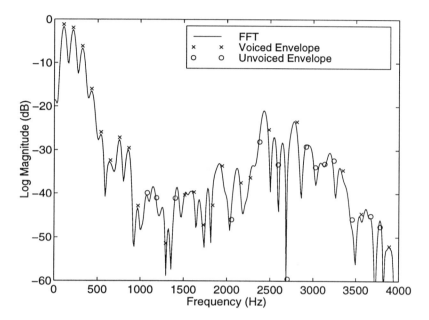

FIGURE 11.2
MBE spectral magnitudes and voicing classifications for a single frame of mixed excitation speech, phoneme /zh/.

11.1.2   Multi-Band Excitation Synthesis

The speech synthesis portion of the MBE model is shown in Figure 11.3. First, the information containing the spectral envelope is separated into voiced and unvoiced sections as dictated by the V/UV bits. The voiced segments contain phase and magnitude information, while the unvoiced segments will contain only magnitude information.

Voiced speech is then synthesized in the time domain by summing sinusoids at harmonics of the fundamental frequency, using the magnitude and phase determined by the voiced envelope information. The magnitude values are linearly interpolated between the previous and current frames to assure smooth transitions.

Unvoiced speech is synthesized from the unvoiced portion of the mag-

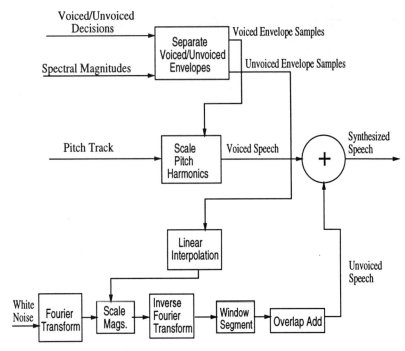

FIGURE 11.3
Speech synthesis in MBE decoder.

nitude stream. A DFT of broadband white noise is amplitude scaled (a different amplitude per channel) so as to resemble the spectral shape of the unvoiced portion of each frame of speech. The spectral amplitudes in the voiced bins are set to zero. An inverse DFT is then applied, each segment is windowed, and the overlap-add method is used to assemble the synthetic unvoiced speech. Finally, the voiced and the unvoiced speech components are added in the time domain to produce the synthesized speech.

Figure 11.4 displays time-domain waveforms for MBE synthesized speech along with the original speech for comparison. The word is "shoes." The initial noise-like, low-energy portion is the /sh/, followed by the high-energy vowel. The mixed excitation /z/ occupies the short final portion. Virtually all of the energy for the vowel is contained in the voiced synthesis component. Conversely, the noise-like fricative /sh/ is synthesized almost entirely with the unvoiced component. The final /z/

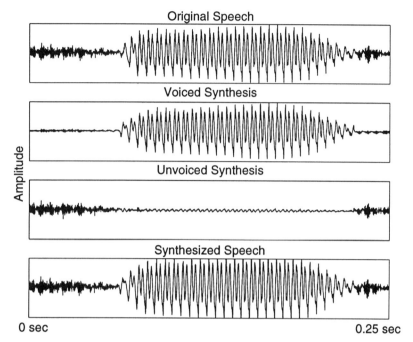

FIGURE 11.4
MBE speech synthesis waveforms for the word "shoes."

has most of its energy in the unvoiced component, but a small periodic contribution can be seen in the voiced component.

11.1.3   Implementations of the MBE Vocoder

A specific implementation of the MBE vocoder was chosen as the standard by the International Mobile Satellite Organization (INMARSAT) for their land mobile satellite standard M [168]. This version, the improved MBE (IMBE), uses 4.15 kbits/s to code the speech parameters, and additional error control coding to raise the total bit rate to 6.4 kbit/s. Further changes to the implementation resulted in the advanced MBE (AMBE). At a bit rate of 3.6 kbit/s for the speech parameters and 4.8 kbit/s rate overall, the AMBE was selected as the INMARSAT Mini-M standard [30] and also the IRIDIUM satellite communications standard. Table 11.1 lists the bit rates for the two MBE implementa-

	Speech (kbit/s)	Error (kbit/s)	Total (kbit/s)	MOS
IMBE	4.15	2.25	6.40	3.4
AMBE	3.60	1.20	4.80	3.7

Table 11.1   Rate and MOS scores of MBE implementations.

tions and the subjective MOS test scores of 3.4 and 3.7 for the IMBE and AMBE, respectively.

References [67], [66], and [96] provide details on the IMBE implementation. A few of the key features are provided here. The frame rate is 20 ms, resulting in 83 bits available to code the speech parameters. The pitch period is linearly quantized with 8 bits over the range of 20 to 115 samples (sampled at 8 kHz) with a quantization step of 1/2 sample. The phases of the voiced harmonics are predicted at the decoder based on the pitch and phase of the previous frame, and the pitch of the current frame. Therefore, no bits are expended on coding the phase. The phases of the voiced harmonics are chosen in such a manner as to maintain coherency across frames.

Each V/UV decision covers a group of 3 harmonics. A maximum of 12 bits (12 bands of 3 harmonics, maximum of 36 harmonics) is allocated to code the voicing information. For cases of low fundamental frequency where the speech frame might contain more than 36 harmonics, the harmonics above the 36th are assumed to be unvoiced for analysis and synthesis.

The IMBE differs primarily in the coding of the magnitude parameters. A prediction scheme operates on the log of the amplitudes. The errors, or residuals, are the difference of the log amplitude of the current frame and the previous frame of the same frequency. The residual amplitudes are grouped into multiple blocks. These blocks are transformed with a discrete cosine transform (DCT) [126]. The zeroth coefficients from the blocks are combined into a "block average vector." The mean is subtracted from the vector. The mean is coded, nonuniformly, with 6 bits. The zero-mean vector is then vector quantized with 10 bit codebook. The remaining DCT coefficients are uniformly scalar quantized with the remaining bits.

11.2   Mixed Excitation Linear Prediction Coder

Mixed Excitation Linear Prediction (MELP) coding [114] is one approach to reduce the bit rate below CELP levels while improving the quality relative to two-state LP coding. MELP employs LPC analysis to model the short-term spectrum, but avoids the hard voiced/unvoiced decision for the entire frame. It models the excitation as a combination of periodic and noise-like components, with their relative contributions based on "voicing strengths" in separate bands across the frequency spectrum. This approach better models segments of speech that have a mixed voicing, as in the voiced fricative /z/, for example, and transitions between voiced and unvoiced.

Figure 11.5 displays the block diagrams for the the MELP encoder and decoder. The coder is similar to a basic two-state LP coder with the added features of mixed excitation, aperiodic pulses, and pulse dispersion filter. These features are discussed below.

Mixed Excitation

As the name implies, the basis for the MELP is a mixed excitation including both a periodic pulse component and a noise-like component. The mixed excitation is designed to reduce what is often claimed as the most annoying quality of LPC synthesized speech, the buzziness of voiced frames [114]. In the general MELP model, the excitation is composed of differing strengths of pulse and noise in each separate frequency band. The number of frequency bands, fixed for all frames in a particular implementation, has been investigated from 4 to 10. The differing strengths are produced at the decoder by the shaping filters shown in the block diagram.

The pulse filter is a sum of the contributions of bandpass filters for each of the frequency bands. Each band contribution is weighted by the voicing strength for that band. The noise filter is generated to have an inverse shape as that of the pulse filter. The noise filter weights are set so as to produce a constant pulse and noise energy in each band. As such, the addition of the filtered pulse and noise components results in an excitation that is spectrally flat.

The voicing strengths are estimated for a particular band based on the bandpass filtered input speech. The voicing strength is computed either as the normalized correlation coefficient of the bandpass speech

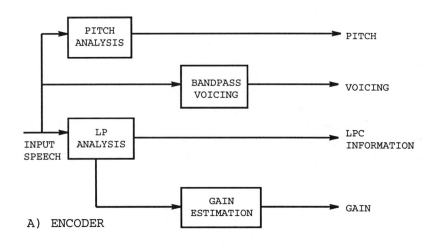

A) ENCODER

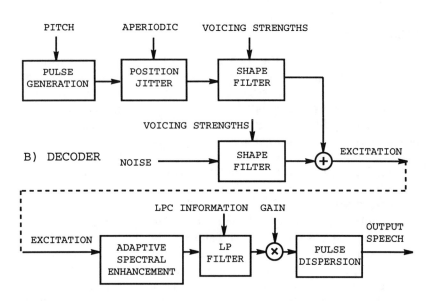

B) DECODER

FIGURE 11.5
Mixed Excitation Linear Predictive (MELP) coder.

at the pitch lag, or as the normalized correlation coefficient of the *envelope* of the bandpass filtered speech. (The envelope is generated by full wave rectification and smoothing with a single pole lowpass filter.) The normalized correlation is defined as:

$$R(\tau) = \frac{\sum_{n=0}^{N-1} s(n)s(n+\tau)}{\sqrt{\sum_{n=0}^{N-1} s^2(n) \sum_{n=0}^{N-1} s^2(n+\tau)}} \tag{11.5}$$

where $s(n)$ is the bandpass filtered speech and N is the frame length. The voicing strength for each band is chosen as the larger of either the correlation coefficient of the speech or the correlation coefficient of the envelope for that band.

Aperiodic Pulses

The aperiodic pulses are designed to remove the LPC synthesis artifact of short, isolated tones in the reconstructed speech. This occurs mainly in areas of marginally voiced speech, when the reconstructed speech is purely periodic. This information is determined at the encoder and passed to the decoder by an "aperiodic flag." The aperiodic flag indicates a jittery voiced state. When the voicing is jittery, the pulse positions are randomized during synthesis based on a uniform distribution around the purely periodic mean position. The pulse can be shifted by as much as $\pm 1/4$ of the pitch period.

The jittery voiced state is set based on the "peakiness" of the full wave rectified LPC residual signal. The peakiness is defined as:

$$peakiness = \frac{\sqrt{\frac{1}{N} \sum_{n=0}^{N-1} r^2(n)}}{\frac{1}{N} \sum_{n=0}^{N-1} |r(n)|} \tag{11.6}$$

where $r(n)$ is the full-wave rectified LP residual and N is the frame length. A threshold is set for the peakiness. In [114], a threshold of 1.8 was suggested, above which, the frame was declared as jittery voiced.

Pulse Dispersion Filter

The pulse dispersion filter aims to produce a better match between original and synthetic speech in regions without a formant by having the signal decay more slowly between pitch pulses. The filter is implemented as a fixed finite impulse response (FIR) filter. The filter is based on a triangle pulse, where the lowpass response has been removed by the process of taking the DFT, flattening the spectral magnitudes, and taking the inverse DFT.

MELP coders typically include an adaptive spectral enhancement filter. This filter has the same design and function as that for CELP coders, and is often referred to as "adaptive postfiltering." The form for the postfilter is that of Equation 10.5 in the section on CELP coding.

In some MELP implementations, the excitation information is augmented by including Fourier coefficients of the LPC residual signal. These coefficients account for the spectral shape of the excitation not modeled by the LPC parameters. These Fourier coefficients are usually estimated from an FFT on the LPC residual signal. The FFT is sampled at harmonics of the pitch frequency. The lower frequency harmonics are considered to be more important and are coded as their difference relative to the mean across frequency. Often, the higher harmonics are not coded explicitly and are assumed to be unity relative to the normalization by the mean value.

11.2.1   Federal Standard MELP Coder at 2.4 kbit/s

In the mid-1990s the U.S. Department of Defense Digital Voice Processing Consortium (DDVPC) tested and selected a new Federal standard to replace the LPC-10e FS1015 at 2.4 kbit/s. The new Federal Standard MELP [115, 156] offers significantly improved speech quality at the 2.4 kbit/s rate. The algorithm closely follows the description in the previous section and includes the Fourier excitation modeling. Particular details of the implementation are presented below.

MELP Encoder

The band edges are shown in Table 11.2. The bandpass voicing analysis operates on 5 frequency bands. The frame rate is 22.5 ms.

Band	Frequency Range (Hz)
0	0 - 500
1	500 - 1000
2	1000 - 2000
3	2000 - 3000
4	3000 - 4000

Table 11.2 Frequency ranges for MELP bandpass voicing analysis.

The initial pitch estimate is based on the maximum of the normalized autocorrelation of 1 kHz lowpass filtered input speech. The allowable range is 40 to 160 samples.

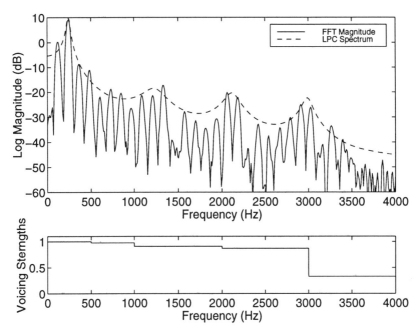

FIGURE 11.6
MELP voicing strengths for a voiced speech frame.

Figures 11.6, 11.7, and 11.8 display the FFT magnitude, LPC spectrum, and voicing strengths for a voiced, unvoiced, and mixed excitation frame, respectively. The voicing strengths are computed for the band structure shown in Table 11.2. The figures plot the unquantized voicing strengths to provide a better indication of the range of the that parameter under the different voicing conditions.

Figure 11.6 shows the spectral plots and voicing strengths for a strongly voiced frame. In the FFT magnitude, the pitch harmonics dominate the frequency range from 0 to 3000 Hz. This is reflected in the voicing strengths for that range. As is common for even strongly voiced speech, the high frequency band of 3000 to 4000 Hz is less periodic and, as such, has a much lower voicing strength.

Figure 11.7 covers the same plots for an unvoiced frame. The FFT magnitude structure is nonperiodic. The voicing strengths are low for all bands. However, the FFT shows some periodic structure for the low band (0 to 500 Hz), and correspondingly, the voicing strength for the low band is higher than the other bands.

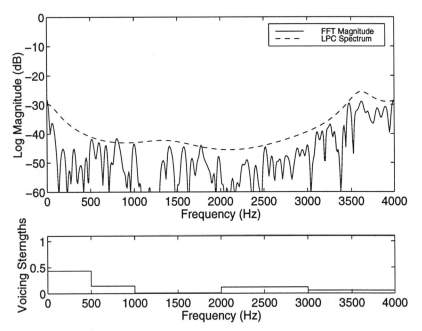

FIGURE 11.7
MELP voicing strengths for an unvoiced speech frame.

Figure 11.8 plots the spectral magnitudes and voicing strengths for a mixed excitation frame. The two low bands, 0 to 500 Hz and 500 to 1000 Hz, indicate a strong periodic structure in the FFT magnitude with the evenly spaced pitch harmonics. The voicing strengths for these bands are high, near 1 for the lowest band. The remaining bands display a less regular, more noisy structure in the FFT magnitude. The voicing strengths for these bands reflect the more noisy, but not entirely random, nature of these bands.

Band 0, the lowest band, is used refine the pitch estimate to subsample accuracy. The normalized correlation coefficient at the fractional pitch estimate for Band 0 is also the voicing strength for Band 0. If that value is less than 0.5, the aperiodic flag is set to true. For the remaining bands, the voicing strength is set to the normalized correlation value for the bandpass signal at the fractional pitch estimate.

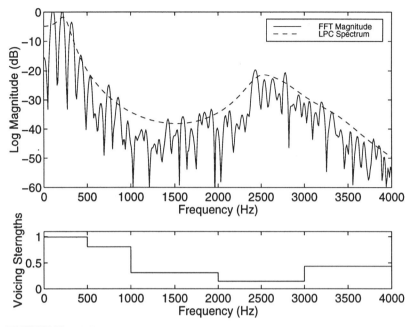

FIGURE 11.8
MELP voicing strengths for a mixed excitation speech frame.

A tenth-order LPC analysis uses a 25 ms Hamming window centered on the last sample of the current frame. The input speech is inverse filtered with the LPC filter to produce the residual. The peakiness value is computed from the LPC residual. If the peakiness value is greater than 1.34, Band 0 voicing is 1.0. If the value exceeds 1.6, Bands 0 through 2 are set to 1.0.

The pitch and Band 0 voicing are quantized together. If Band 0 voicing is less than 0.6, the frame is classified as unvoiced, and all band voicings are set to zero. If voiced, the log of the pitch is uniformly quantized to 100 levels. For the voiced case, Bands 1 to 4 are quantized to 1 if their voicing strength is greater than 0.6; otherwise, they are set to 0.

The LPC parameters are converted to line spectral frequencies (LSFs). The LSFs are quantized with a 4-stage vector quantization (VQ) algorithm. The first stage has 7 bits, while the other three have 6 bits each. The resulting quantized vector is the sum of the vectors from each stage, one per stage. At each stage in the search phase, the VQ search finds the

"M best" closest matches to the original. In this standard, M is equal to 8. These M best are used in the search for the next stage. The indices of the final best at each of the four stages determine the quantized LSF.

The Fourier magnitudes are estimated from an FFT of the LPC residual for voiced frames only. The magnitudes of the first 10 pitch harmonics are selected. They are coded with an 8-bit VQ using a perceptually weighted distance measure.

Several parameters are not transmitted for unvoiced frames, including the Fourier magnitude, the aperiodic flag, and the voicing for Bands 1 to 4. These 13 bits are used for error correction coding.

Table 11.3 details the bit allocation for voiced and unvoiced frames.

Parameter	Voiced	Unvoiced
LSF	25	25
Fourier	8	0
Gain	8	8
Pitch, Band 0	7	7
Bands 1 - 4	4	0
Aperiodic flag	1	0
Error coding	0	13
Synchronization	1	1
Total	54	54

Table 11.3   Bit allocation for MELP standard coder [156].

MELP Decoder

The pitch is decoded first because it contains the information about the voicing state and whether error coding was used. The pulse excitation results from an inverse DFT of one pitch period. The noise and pulse components are filtered by the bandpass information for each band, and the two components are added in the time domain.

The excitation is enhanced by the adaptive postfilter, multiplied by the gain, and filtered by the LPC coefficients. Finally, the signal is passed through the pulse dispersion filter to produce the output speech.

Figure 11.9 displays the original time waveform, the voiced (pulse) and unvoiced (noise) synthesis components, and the added total synthesis. The speech segment is the word "shoes." The unvoiced fricative /sh/ begins the segment, and the major central portion is the vowel. The final consonant /z/ is formed by mixed excitation. The unvoiced /sh/

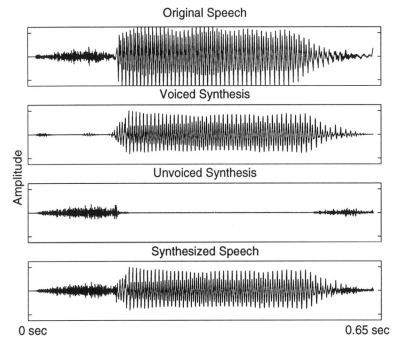

FIGURE 11.9
MELP synthesis waveforms for the word "shoes."

is synthesized almost entirely by the unvoiced component. The vowel is composed of voiced synthesis only. The voiced fricative /z/ clearly shows the mixed excitation nature of this sound. The periodic pitch pulses are evident in the voiced synthesis portion of this phoneme. However, the /z/ has a strong noise component also, as can be seen in the final portion of the unvoiced synthesis waveform.

MELP Performance

Reference [95] has taken an in-depth look at the performance of the Federal Standard MELP coder compared with the FS1015 LPC-10e, the FS1016 CELP, and a 16 kbit/s continuously variable slope delta modulator (CVSD). The coders were compared under a variety of background noise conditions and simulated transmission bit errors.

The MELP algorithm performed as well as the CELP algorithm (4.8 kbit/s). The MELP achieved a MOS score 3.3 with a quiet background, compared with 3.1 for CELP and 2.2 for LPC-10e.

Appendix A includes on-line locations that contain both examples of speech coded and decoded by LPC-10e and downloadable source code in Fortran or C of a software implementation.

11.2.2 Improvements to MELP Coder

With its selection as the new Federal standard at 2.4 kbit/s by the U.S. Department of Defense, the MELP scheme has been the focus of much experimentation. Efforts have been directed at improving the quality at the 2.4 kbit/s rate [160], reducing the bit rate to 1.7 kbit/s [116], and attempting to achieve near toll quality at 4.0 kbit/s [154].

Improvements to 2.4 kbit/s Coder

The improvements suggested by Unno, Barnwell, and Truong in [160] pertain to the 2.4 kbit/s MELP and maintain the bit stream specification of the Federal standard. They have implemented three quality improvements:

- Improved pitch estimation to reduce artifacts in transition segments

- Plosive detection and specialized synthesis

- Post processing of Fourier magnitudes for low pitch male speakers

The improved pitch estimation attempts to better follow the pitch track at the ends of vowel segments. The method incorporates a sliding pitch window. The window for the normalized autocorrelation pitch estimate is moved forward and backward around the centered position. The position that achieves the highest correlation value is chosen, along with that corresponding pitch period.

The peakiness of the LPC residual is used to locate plosives (stop consonants, such as /p/ or /g/). A sliding window is used to locate the position where the peakiness is a maximum. To separate plosives from vowel onsets, the lowpass energy is examined (vowel onsets will have significantly more energy than plosives). The plosive is modeled and synthesized by a single LP residual signal. The LP residual is multiplied by a gain and filtered by the LP parameters for the frame.

The first few harmonics for low pitch male speakers are boosted to account for an attenuation due to the adaptive spectral enhancement filter and a 60 Hz highpass filter. The equalization to compensate these

Parameter	FS 2.4 kbit/s	1.7 kbit/s
LSF	25	21
Fourier	8	0
Gain	8	5
Pitch, Band 0	7	6
Bands 1−4	4	2
Aperiodic flag	1	0
Synchronization	1	0
Total	54	34

Table 11.4 Bit allocation comparison between 2.4 kbit/s MELP standard and 1.7 kbit/s MELP[116]

effects is applied to the Fourier magnitudes located 200 Hz below the first formant frequency. The first formant is estimated roughly based on the LSP coefficients.

Reduction of Bit Rate to 1.7 kbit/s

The work presented in [116] was directed at reducing the bit rate below 2.4 kbit/s. The paper reports improved quality over the Federal Standard with a lower bit rate at 1.7 kbit/s. Improvements in pitch and voicing estimation, noise suppression front-end processing, and a reduction of the frame rate from 22.5 to 20 ms were cited for the quality improvements. A 21-bit switched predictive quantization of the LSP parameters reduces the bit rate from the Federal Standard (25 bits) while achieving lower spectral distortion. In the switched predictive method, two separate predictors and codebooks are trained and used for quantization. In quantization, the predictor−codebook pair that yields lower distortion is selected. One bit is required to code which pair was used. The Fourier magnitudes are not transmitted, and reducing and rearranging the bit allocation facilitates the 1.7 kbit/s rate. Table 11.4 compares the bit distributions between the two coders.

High Quality 4.0 kbit/s Coder

The efforts of [154] were aimed at producing high quality speech at 4.0 kbit/s with a MELP coder. The LSF quantization is changed to a switched predictive scheme similar to [116]. Also, the Fourier magnitudes are coded with a switched predictive VQ. Additional bits are used to quantize information about the interpolation paths of the LSFs across the 20 ms frame. The gain and pitch are computed twice each frame, and the voicing information (5 bands) is also computed and transmitted two times each frame. In summary, most of the additional bits are used to code the Fourier magnitudes more accurately, and to update the pitch, gain, and voicing information twice per frame.

The performance of the coder was evaluated in subjective listening tests. The coder was compared in an A/B pairwise paradigm. For clean input speech, the 4 kbit/s MELP was found to be better than the GSM Full Rate 13 kbit/s standard, but not as good as the ITU G.729 CS-ACELP at 8 kbit/s. The 4 kbit/s MELP was nearly equal to the quality of the G.729 in the presence of background car noise [154].

11.3   Split Band LPC Coder

Split Band LPC coding [6, 172] (also termed Harmonic Excitation LPC) is similar to MELP coding in that it is designed as a low rate coder with LP modeling, a mixed excitation model, and separate voiced and unvoiced synthesis. However, the Split Band algorithm sections the excitation spectrum into two frequency bands with a variable dividing frequency. The lower band represents the voiced excitation and the higher band models the unvoiced portion. Makhoul had suggested a similar concept of split band, voiced/unvoiced LP excitation in [106].

The Split Band LP method is also similar to, and has its basis in, MBE coding. It can be interpreted as an MBE method where the spectral magnitudes are represented by the combined parameters of the LP spectrum and the residual magnitudes. In the Split Band, the multiband voicing of the MBE is quantized into two bands (lower voiced, upper unvoiced) of varying complementary bandwidths, covering the whole spectrum.

The general algorithm of the coder is similar to MELP. An LP analysis is performed on the input speech. The LP coefficients are converted to LSFs and vector quantized by an appropriate scheme. The initial pitch estimates are refined to higher accuracy. The harmonic magnitudes are estimated by sampling the magnitude of the FFT of the LP residual. The algorithm represents the voicing information in a manner different from MELP.

The voicing information is referred to as the voicing probability because of the method of computation. The voicing probability determines the dividing frequency between the voiced and unvoiced portions of the spectrum. Several techniques have been suggested for its computation [6, 172, 171].

In [172], the voicing probability is computed by first determining the voiced/unvoiced decision for each harmonic of the pitch. This is done by constructing a synthetic spectrum by assuming it is completely voiced. This synthetic is compared to the original for each harmonic. If the match is good, that harmonic is classified as voiced. This is the same method as described for the MBE.

Given the V/UV decisions for each harmonic, the voicing probability is computed as the ratio of the energy in the voiced harmonics to the total energy as:

$$Pr_{voicing} = \sqrt{\frac{\sum_{k=1}^{K} V(k)S^2(k)}{\sum_{k=1}^{K} S^2(k)}} \tag{11.7}$$

where $S(k)$ is the spectral amplitude at the kth harmonic, $V(k)$ is the voicing decision for the kth harmonic, and K is the number of harmonics in the total bandwidth. $V(k) = 1$ for voiced harmonics, and $V(k) = 0$ for unvoiced.

The result is a percentage between 0 and 1 that corresponds to a percentage of the frequency band (0 to 4 kHz) where the split between voiced and unvoiced occurs. This frequency was quantized with 3 bits in both [172] and [6].

11.3.1 Bit Allocations and Quality Results

Two separate Split Band vocoder bit allocations and performances are shown here. Reference [6] presented a 2.5 kbit/s Split Band coder with the bit allocation as displayed in Table 11.5.

For this 2.5 kbit/s implementation, subjective listening tests compared the Split Band coder to the IMBE and the FS1016 CELP. The overall MOS results are shown in Table 11.6 for a mix of male and female speakers under clean conditions. The results indicate that the Split Band coder, at a much lower bit rate, performs as well as or slightly better than the IMBE and Federal Standard CELP.

The 4 kbit/s Split Band coder reported in [172] applies a 14th order LP analysis to the input speech. The LSFs are split into four groupings

Parameter	2.5 kbit/s
LSF	28
Harmonic Mag	6
Gain	6
Pitch	7
Voicing	3
Total	50

Table 11.5 Bit allocation of 2.5 kbit/s Split Band LPC [6].

Coder	Mean Opinion Score
IMBE 4.15 kbit/s	3.2
FS1016 CELP 4.8 kbit/s	3.1
Split Band 2.5 kbit/s	3.4

Table 11.6 Comparison of MOS results for 2.5 kbit/s Split Band [6].

(3, 3, 4, 4) and the log values are transformed with the DCT. The DCT coefficients are vector quantized with 10 bits for each grouping. The residual harmonics are also vector quantized in the DCT domain. For both the LSFs and the residual harmonics, the first subframe is approximated with a linear interpolation. The information about this interpolation is quantized to 3 bits. The pitch for the second subframe is quantized with 7 bits, while the first subframe uses the differential pitch, quantized with 5 bits. As previously mentioned, the voicing dividing frequency is coded with 3 bits. Table 11.7 lists the bit allocation for the 4 kbit/s Split Band coder.

In subjective listening tests, the 4.0 kbit/s Split Band coder was com-

Parameter	4.0 kbit/s
LSF	3, 40
Harmonic Mag	3, 19
Pitch	5, 7
Voicing	3
Total	80

Table 11.7 Bit allocation of 4.0 kbit/s Split Band LPC [172].

Coder	Mean Opinion Score
AMBE Mini-M 3.6 kbit/s	3.35
IS-54 VSELP 8.0 kbit/s	3.66
G.729 ACELP 8.0 kbit/s	3.40
Split Band 4.0 kbit/s	3.68

Table 11.8 Comparison of MOS results for 4.0 kbit/s Split Band [172].

pared to the IS-54 Vector Sum Excited Linear Prediction (VSELP) standard at 8 kbit/s, the ITU G.729, and the INMARSAT AMBE Mini-M at 3.6 kbit/s. MOS data suggest that the Split Band offers comparable performance to the 8 kbit/s VSELP and slightly better performance that the other two coders. The MOS test results are listed in Table 11.8.

11.4 Harmonic Vector Excitation Coder

Harmonic Vector Excitation (HVXC) coding [121, 185] is part of the MPEG-4 audio coding standard, and is used to code narrowband (8 kHz sampling rate) speech at 2.0 or 4.0 kbits/s. The coder also supports a variable rate mode at 1.2 to 1.7 kbit/s. The coder is LP based, vector quantizes the spectral shape of the LP residual for voiced frames, and employs a CELP scheme (also referred to as vector excitation (VXC)) to encode the LP residual for unvoiced frames. The application of CELP for unvoiced speech differentiates the HVXC from most other low rate coders, which usually synthesize the unvoiced excitation with random noise, instead of fitting codebook entries.

11.4.1 HVXC Encoder

Figure 11.10 displays a simplified block diagram for the encoder. The frame rate is 20 ms. A tenth-order LP analysis is performed on the input speech, and the LSPs are computed once per frame. The LSPs are quantized at 18 bits for the 2.0 kbit/s coder and 26 bits for the 4.0 kbit/s version. The 2.0 kbit/s version is referred to as the base layer, and the 4.0 kbit/s version as the enhanced. The VQ is 2-stage for the

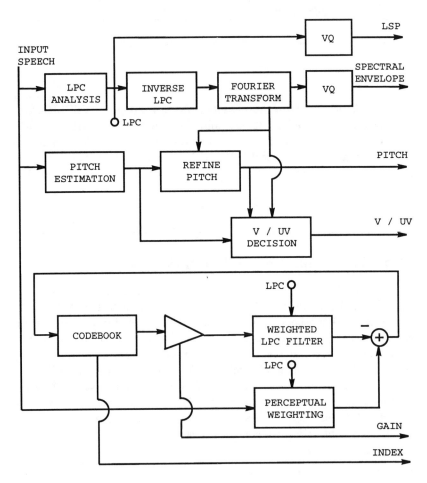

FIGURE 11.10
Harmonic Vector Excitation (HVXC) encoder.

base layer. The first stage quantizes the 10 LSPs with 5 bits. Multiple candidates from the first stage are evaluated at the second stage to find the lowest overall error. The second stage splits the error from the first stage into two 5-element vectors. The first 5 are quantized with 7 bits, and the last 5 with 5 bits. One additional bit is used to signify whether interframe prediction resulted in an overall lower error. The enhanced version vector quantizes the error from the first two stages of the base layer (yielding three stages overall) with a single 10-element codebook of 8 bits. This coding scheme for the LSPs is the same for voiced and unvoiced frames.

The autocorrelation of the LP residual is used for the initial pitch estimate. The pitch estimate is refined by the manner described in the MBE coder of Section 11.1.1. A small range of fractional lag values around the initial estimate is searched for the one that produces a synthesized spectrum that best matches the original spectrum. If the frame is voiced, the pitch value is coded using 7 bits.

The spectral envelope is estimated by the DFT magnitudes at the harmonics of the pitch frequency. The magnitudes are transformed to a fixed dimension (44) by bandlimited interpolation because the original dimension will vary based on the number of pitch harmonics in the 0 to 4000 kHz band. The base version employs a 2-stage VQ to encode the spectral magnitude information. The base version uses 4 bits at each stage for the vector shape, and 5 bits for the gain. For the enhanced layer, the error from the base version is further vector quantized with a 4-split VQ, with an additional 32 bits.

The V/UV decision is based on the match between the synthesized and original spectrum from the pitch refinement, the peak of the autocorrelation of the LP residual (normalized by the residual power), and the number of zero crossings of the time-domain waveform. High rate of zero crossings is indicative of noise-like unvoiced speech. The V/UV information is divided into four classes: unvoiced, background noise, mixed voicing, and voiced. These classes are important for scaling the spectral magnitudes during synthesis (discussed in the next section).

When a frame is classified as unvoiced, the lower signal flow of Figure 11.10 encodes the unvoiced frames by the CELP method. The CELP encoding is the same as discussed in Section 10.4, without the long term predictor (LTP) because the speech is known to be unvoiced. For the base version, 6 bits code the shape and 4 bits code the gain, twice per frame (each 10 ms). For the enhanced version, 5 shape bits and 3 gain bits describe the excitation four times per frame (each 5 ms).

The variable rate version reduces the bit rate of the base version by detecting the background noise condition. When no speech is present, 1 out of 9 frames is sent with unvoiced information to coarsely approximate the background noise. The other 8 frames just indicate the "background noise" state with the 2 V/UV bits. To further reduce the bit rate for unvoiced frames containing actual speech, the excitation is encoded with only 8 bits of gain information.

11.4.2   HVXC Decoder

Figure 11.11 illustrates the block diagram for the HVXC decoder. The LSP vector is reconstructed from the 2-stage VQ (where the second stage is split, 5/5) for the base version. For the enhanced layer, the third stage vector is added onto the result of the previous two stages. The LSPs are stabilized by arranging them in increasing order and assuring a minimum spacing.

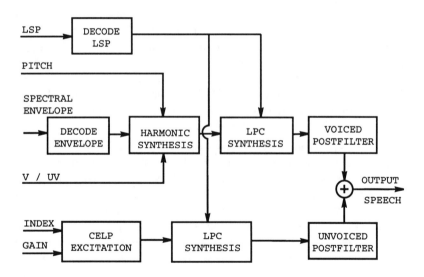

FIGURE 11.11
Harmonic Vector Excitation (HVXC) decoder.

For frames classified as unvoiced, the CELP codebook entry is multiplied by the gain to generate the unvoiced excitation. The unvoiced excitation is filtered by the LPC synthesis filter, and that result is fil-

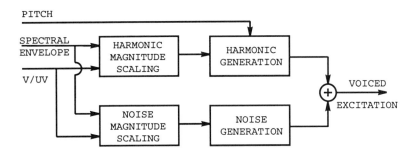

FIGURE 11.12
Synthesis of voiced excitation: Combination of harmonic and noise components, weighted by scaled spectral magnitudes.

tered by the unvoiced postfilter. The unvoiced postfilter is of the form of Equation 10.5, where $\mu = 0.1$, $\beta = 0.5$, and $\alpha = 0.8$.

For all frames not unvoiced, the spectral magnitudes are formed by combining the vectors from the stages as described in the encoding. The fixed-length magnitude vector is transformed to the appropriate variable length (based on the pitch) by bandlimited interpolation. The "Harmonic Synthesis" block generates the voiced excitation as the sum of the weighted pitch harmonics plus spectrally weighted noise. These weightings depend on the V/UV classification. A more detailed block diagram of the "Harmonic Synthesis" block is shown in Figure 11.12. The spectral magnitudes are scaled and used for harmonic generation and noise generation. The noise generation takes the DFT of time-domain white noise, shapes the spectrum based on the noise scaled magnitudes, and performs an inverse DFT. The harmonic and noise components are added in the time domain.

The harmonic and noise scalings of the spectral magnitudes depend on frequency and V/UV decision. Specifically, Figure 11.13 displays the scale factors. The noise magnitudes are zero at low frequencies and are a fraction of the original spectral magnitude at high frequencies. The mixed classification has a significant noise contribution at high frequencies (from 0.85×4000 Hz to 1.0×4000 Hz) of 0.5 of the original spectral magnitudes. Correspondingly, the harmonic magnitude is reduced to 0.5 its original value.

The voiced excitation is filtered by the LPC synthesis filter, then that output is filtered by the voiced postfilter. The voiced postfilter is the same form as the unvoiced postfilter and Equation 10.5 with one small change. The term μ depends on a_1, such that $\mu = -0.15a_1$, with the further limitation that $0 \leq \mu \leq 0.5$. The values of β and α are the same as the unvoiced postfilter, $\beta = 0.5$, and $\alpha = 0.8$. The voiced and unvoiced synthesis components are added in the time domain to produce the final synthesized speech.

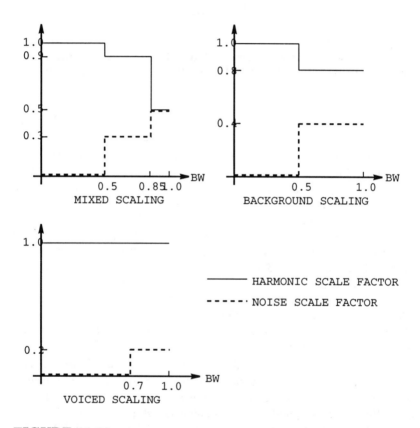

FIGURE 11.13
Scale factors used to weight the harmonic magnitudes for voiced excitation synthesis. X-axis scale is the bandwidth of the signal where 1.0 corresponds to 4000 Hz.

Coder	Mean Opinion Score
HVXC 2.0 kbit/s	2.53
HVXC 4.0 kbit/s	2.92
FS1016 CELP 4.8 kbit/s	2.19

Table 11.9 Subjective listening test comparing HVXC coder to Federal Standard 1016 [184].

11.4.3 HVXC Performance

The performance of the 2.0 and 4.0 kbit/s HVXC coders was compared to the 4.8 kbit/s U.S. Federal Standard 1016 (FS1016) CELP coder and reported in [184]. The test was conducted as an absolute category rating, or mean opinion score. The categories ranged from 5 = "Excellent," 4 = "Good," 3 = "Fair," 2 = "Poor," down to 1 = "Bad." The test material included German, English, Swedish, and Japanese speech samples.

The means of the test ratings are shown in Table 11.9. Both the 2.0 and 4.0 kbit/s HVXC coders provide higher quality than the FS1016 standard.

11.5 Waveform Interpolation Coding

Waveform Interpolation (WI) [92, 146, 94] was originally conceived as a method to efficiently encode a pitch period of voiced speech. For voiced speech, the shape of the pitch period waveform changes slowly from one period to the next. Because of the slow change over time, the representation of the waveform can be significantly downsampled for efficient coding. Upon decoding, the intermediate representations are interpolated. This is possible because of the smooth, slow changes.

In current implementations of WI, an LP analysis and filtering remove the vocal tract frequency shaping. The residual is represented as a "characteristic waveform" and extracted at least once each pitch period for both voiced and unvoiced speech. For purposes of decomposition and coding gain, the residual is characterized by a two-dimensional signal usually referred to as $u(t, \phi)$. At a particular sampling time, t_i, the shape of the residual is represented along the ϕ axis. The variable u is periodic along ϕ and in most implementations is specified as a Fourier series. Along the t axis, u changes as the shape of the residual changes at different sampling times, t_{i-1}, t_i, t_{i+1}. To restate, along the ϕ axis, $u(t_i, \phi)$ contains a Fourier series of a segment of the residual signal, centered near t_i.

The signal u changes, or evolves, slowly in time (along t) for voiced speech and rapidly for unvoiced speech. This results directly from the residual signal being very similar from one pitch period to the next for most voice speech, and random for unvoiced. The signal $u(t, \phi)$ can be separated into a "slowly evolving waveform" (SEW) and a "rapidly evolving waveform" (REW). The SEW corresponds to the periodic, voiced component of the residual. The unvoiced, noisy portion is represented as the REW. The separation of the SEW and REW is carried out by filtering on the t axis. Lowpass filtering along the t axis yields the SEW, $u_{SEW}(t, \phi)$. Highpass filtering u along the the t axis results in $u_{REW}(t, \phi)$. The cutoff frequency for both the highpass and lowpass filters is the same so that:

$$u(t, \phi) = u_{SEW}(t, \phi) + u_{REW}(t, \phi) \qquad (11.8)$$

The basic concept of the WI coder is discussed in the next section, while the quantization of the REW and SEW and accompanying coding gain are outlined in Section 11.5.2.

11.5.1 WI Coder and Decoder

Figure 11.14 displays the encoder and decoder diagrams for WI, minus the encoding and decoding of the REW and SEW. An LP analysis is performed on the input speech. The input speech is filtered by the inverse LP filter to produce the residual. The pitch is estimated from the residual signal. Given the residual and the pitch period, the characteristic waveform is segmented from the residual with a rectangular window. Some latitude is given to allow placement of the ends of the rectangular window near low values of the residual signal to reduce discontinuities at the endpoints. The characteristic waveform is aligned with the last previously extracted characteristic waveform. The characteristic waveform is converted to the Fourier domain. While the sequence of processing steps described here is easier to visualize, in practice, the alignment operation is performed after conversion to the Fourier series representation.

Filtering the characteristic waveforms along the t axis separates the REW and SEW. The output of the lowpass filter yields the SEW, and the

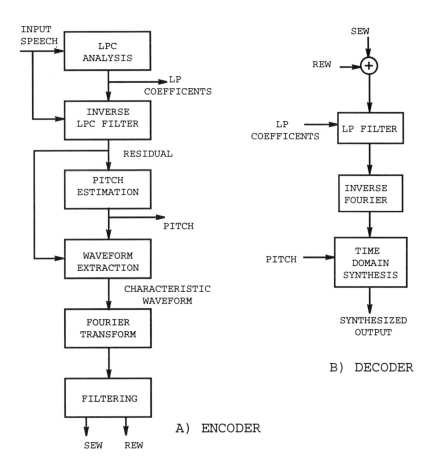

FIGURE 11.14
Waveform Interpolation (WI) encoder and decoder.

output of the highpass filter, the REW. In practice, a cutoff frequency of 20 Hz for both filters has given the desired separation.

At the decoder, the REW and SEW are added to obtain the Fourier domain version of the characteristic waveform. Vocal tract frequency shaping of the LP coefficients is carried out by filtering in the frequency domain. The vocal-tract-shaped Fourier-domain signal is inverse transformed to produce the two-dimensional speech-domain surface. The time-domain output speech is synthesized based on the pitch values. Further details can be found in [94].

11.5.2   Quantization of SEW and REW

The entire basis for WI coding is the separation of the REW and SEW to allow efficient quantization of each individually. The SEW requires accurate quantization for low frequency components because low frequencies are more perceptually important. The accurate, low frequency SEW representation need only be updated at a low rate (typically once per 25 ms frame) because of its slowly varying nature. For the REW, only the general shape of the magnitude spectrum is encoded with a course quantization. The REW is updated at a higher rate of 2 to 4 times per frame.

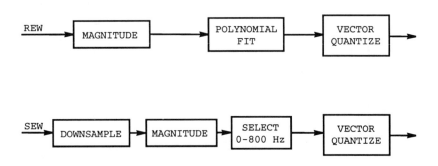

FIGURE 11.15
Quantization of rapidly evolving waveform (REW) and slowly evolving waveform (SEW) for WI encoder.

Figure 11.15 highlights the quantization of the REW and SEW. From the Fourier series REW, the magnitude is computed. Polynomial coefficients are fit to the magnitude spectrum. The polynomial is used to represent the varying length Fourier series magnitude as a fixed length vector. The low-order (fifth-order in reference [94]) polynomial characterizes only the slope and broad shape of the spectrum. In several implementations of the WI coder, the polynomial coefficients are vector quantized using only 8 shapes (3 bits).

The SEW is downsampled to the frame rate. The magnitude of the spectrum from 0 to 800 Hz is selected. The selected low frequency segment is vector quantized using about 7 or 8 bits, once per frame. The phase information for both the REW and SEW is not transmitted; their structure is assumed at the decoder.

At the decoder, the general shape of the REW magnitude is recovered from the quantized polynomial coefficients. The REW phase is assumed to be random because of the noise-like qualities of the mostly unvoiced residual from which it is derived. The low frequency portion (0 to 800 Hz) of the SEW magnitude is recovered from the quantized SEW. The overall magnitude spectrum of the characteristic waveform (and residual) is assumed to be flat. As such, the higher frequency portion (800 to 4000 Hz) of the SEW is forced to fit the equation:

$$|SEW| \approx 1 - |REW| \qquad (11.9)$$

The phase of the SEW is approximated at the decoder as a fixed linear phase taken from an example segment of voiced speech.

11.5.3 Performance and Enhancements

Kleijn and Haagen reported on a 2.4 kbit/s WI coder in [93]. The bit allocation and parameter update rates are shown in Table 11.10. The frame rate is 40 Hz (25 ms). The LP coefficients are quantized with a split VQ of the LSFs. The log of the signal power is differentially quantized twice per frame. The REW magnitude is updated 6 times per frame. For the first, third, and fifth updates, the magnitude is vector quantized with 3 bits. For the intermediate updates, a single bit selects either the previous or following REW magnitude, whichever is a closer fit.

The 2.4 kbit/s WI coder was compared to the FS1016 CELP at 4.8 kbit/s under a variety of testing conditions. The WI coder was equivalent to or better than the FS1016 under all tested conditions. A few of the test results for quiet background noise conditions are displayed in Table 11.11.

The early version of the WI coder required a computational complexity far beyond what could be reasonably implemented for real time oper-

Parameter	Update Rate (per frame)	Bits
LP	1	30
pitch	1	7
power	2	2x2
REW magnitude	6	3x3, 3x1
SEW magnitude	1	7

Table 11.10   Bit allocation for Waveform Interpolation coder of [93].

Test	WI	FS1016
MOS	3.77	3.59
DAM	66.8	63.1
DRT	87.2	87.7

Table 11.11   Subsection of tests results comparing 2.4 kbit/s WI coder of [93] to FS1016 4.8 kbit/s CELP.

ation. In [94], complexity reductions lowered the required computations by a factor of 10, while maintaining the speech quality.

In [87], Kang and Sen added improvements to the basic WI approach by modifying the REW spectrum based on the pitch. The modification reduces the REW contribution, and corresponding noisy sound, for certain speech segments where the pitch changes rapidly. The REW is modified in different manners depending on whether the speaker is male or female. Because the male/female decision and location of areas of rapidly changing pitch can be determined at the decoder based on the decoded pitch track, no additional information needs to be transmitted. The bit rate remains at 2.4 kbit/s. In A/B comparison tests, the new coder was preferred over the conventional WI by a wide margin.

Recently, Gottesman and Gersho presented a 4.0 kbit/s WI coder with several enhancements [58]. The enhancements include an analysis-by-synthesis SEW search for the best quantized vector, an analysis-by-synthesis quantization of the phase of the SEW, an improved pitch search for transition regions, and a switched-predictive gain vector quantization. The bit allocations for the 4.0 kbit/s coder are displayed in Table 11.12.

In separate subjective A/B comparison tests, the 4.0 kbit/s WI coder was preferred by a wide margin over both the MPEG-4 4.0 kbit/s HVXC coder and the 5.3 kbit/s G.723.1 ACELP. The WI coder was favored

Parameter	Bits/Frame
LP	18
Pitch	2x6
Gain	2x6
REW	20
SEW mag.	14
SEW phase	4

Table 11.12   Bit allocation for 4.0 kbit/s WI coder of [58].

WI 4.0 kbit/s	4.0 kbit/s MPEG-4
63.7%	36.3%
WI 4.0 kbit/s	5.3 kbit/s G.723.1
59.5%	40.5%
WI 4.0 kbit/s	6.3 kbit/s G.723.1
53.9%	46.1%

Table 11.13   Subjective A/B comparison listening tests for 4.0 kbit/s WI coder of [58] relative to standard coders.

slightly over the 6.3 kbit/s G.723.1 MP-MLQ coder. The comparison tests were conducted with clean input speech, and included both male and female speech. The test results are listed in Table 11.13.

Chapter 12

Perceptual Speech Coding

The goal of perceptual coding is to reduce the size of the signal representation while maintaining the perceived sound quality by exploiting the limits of human auditory perception. The exploitable limits of auditory perception stem from the frequency and temporal masking discussed in Chapter 6.

This chapter presents an overview of perceptual speech coding. Frequency and temporal masking are considered together to determine which signal components are not perceptible. The impact of the sound quality (tone or noise) of the maskee and masker is discussed. Because of the limited time and frequency resolution of standard frequency-domain transforms (discrete Fourier transform), the Multi-Band Excitation (MBE) speech model is shown to have advantages for perceptual coding. The last section lists a sampling of the current research in perceptual speech coding.

While this chapter discusses perceptual speech coding schemes, these approaches are not as dominant as perceptual approaches for general wideband audio coding. To date, the highest quality, lowest rate speech coders are of the type described in Chapter 11. However, the progress of future research holds the promise of perceptual coding gains for speech.

12.1 Auditory Processing of Speech

Section 6.4 discussed monaural masking. One sound can mask another simultaneous, lower amplitude sound when the two are close in frequency. This is referred to as "simultaneous masking in frequency."

When two sounds occur at nearly the same time, the lower level signal can be masked by the stronger signal in the phenomena of "temporal masking." The challenge of perceptual coding is how to determine which sounds mask which other sounds in a complex, rapidly varying speech signal.

12.1.1   General Perceptual Speech Coder

Most of the algorithm processing steps of a perceptual speech coder are similar to those of conventional speech coders. The primary difference is the determination and deletion of signal components that are not perceptible.

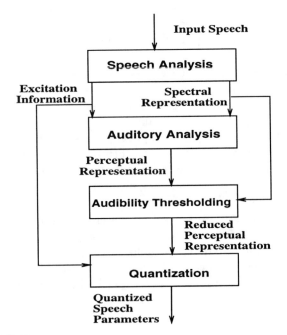

FIGURE 12.1
General perceptual speech coder.

Figure 12.1 displays a block diagram of a general perceptual speech coder. The input speech is analyzed, yielding a short-term spectral representation of the vocal tract and excitation information. These parameterizations are transformed by an auditory analysis into a perceptual representation. In the perceptual representation, the frequency scale is warped to a nonlinear scale based on critical bands (see Chapter 6).

Within the perceptual domain, masking and masked signal components are determined. The masked components, which are not perceptible, are deleted from the representation or marked to be coded more coarsely, that is, with fewer bits. The reduced perceptual representation results in a lower bit rate due to the reduced number of parameters, but is used to synthesize output speech of the same perceived quality as the complete representation. Determining the particulars of the masking is discussed in more detail in the following sections.

As with any speech encoding/decoding system, the decoder merely reverses the operations to synthesize the output speech.

12.1.2 Frequency and Temporal Masking

It is well known that simultaneous masking in frequency is more prominent when the masker is lower in frequency than the maskee. Referring back to Figure 6.4, the plotted threshold of detectability is much lower for frequencies below the masker, than for frequencies above.

This observation suggests an efficient method to determine which components are masked:

1. Transform each short time segment of speech into the frequency domain.

2. Segment frequency domain representation into logarithmically spaced frequency bands (constant number of barks per frequency segment).

3. Calculate the total energy in the lowest band.

4. Determine the threshold of detectability within this critical band and in the higher frequency critical bands.

5. Code only frequency information above the threshold level.

6. Continue threshold calculation/coding process for the next higher critical band.

7. Repeat steps 3 through 6 until all critical bands are coded.

Although more complex, this method could be extended to include masking regions where the maskee is in a lower frequency critical band

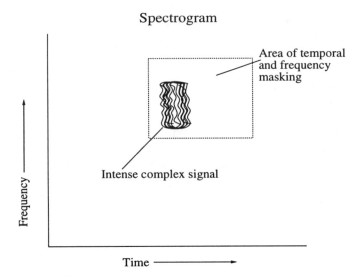

FIGURE 12.2
Island of perceptually significant signal and resulting area of masking.

than the masker. The previously described method is highly efficient; however, it does not take full advantage of the properties of simultaneous masking. The method calculates a saturation threshold level for each critical band, but it does not take into account spectral changes within each critical band. The described method does not consider the effect of temporal masking across different frames.

Simultaneous frequency and temporal masking suggest that substantial economies in coding can be gained by spectral analysis to determine "islands" of perceptually significant signals in the time/frequency/intensity dimensional representation. Figure 12.2 shows an intense complex signal surrounded in the time/frequency domain by a box which represents the signals that would be masked in the presence of this complex signal. These complex signals appear as high intensity "islands" in a typical spectrogram. The majority of the available coding capacity can then be assigned to accurately represent these islands and a minimum assigned to regions masked by these islands.

12.1.3   Determining Masking Levels

In Figure 12.1, an auditory analysis of the speech parameters is performed on each frame of speech. The auditory analysis transforms the signal representation into the perceptual representation. The high intensity regions in the time-frequency plane mask (either somewhat or completely) some of the less intense regions as in Figure 12.2. This masking causes the threshold of detectability of the maskee to be increased. If the threshold of detectability of a region is greater than the intensity of that region, then the portion of the signal denoted by that region is not audible to human hearing. These values are calculated by comparing all the regions to each other and determining how much the threshold of detectability is raised for each time-frequency region. Psycho-acoustic data such as those represented in Figures 6.4 and 6.5 are used in the calculations of these values.

Figure 12.3 is a 3-dimensional representation of the union of a particular set of simultaneous and temporal masking data. The time scale is the time difference between the masker and maskee, and the frequency scale is the frequency difference between the two. The peak of the surface is the origin, where the relative time, relative frequency, and relative amplitude are all zero.

This graphical representation can lend insight into the workings of perceptual speech coding. A time/frequency/magnitude representation of a speech utterance can be displayed as a 3-dimensional surface. This is a 3-D representation of the spectrograms of Chapter 2, where the amplitude, displayed as shades of gray in the spectrograms, is now the vertical height of the surface.

Figure 12.4 displays this data representation for a half-second segment of speech for the frequency range of 0 to 1000 Hz. This representation can be visualized as a mountainous landscape. High elevation areas correspond to high amplitude signals regions located at particular time and frequency coordinates. The ridges running across time, of nearly constant frequency, are the pitch harmonics. (The same pitch harmonics appear as dark bands in the spectrograms of Chapter 2.) If the mountainous speech landscape is divided up into segments, the time divisions correspond to different analysis frames, and the frequency divisions correspond to dividing the spectrum into critical bands.

Visualize a copy of Figure 12.3 (appropriate for the frequency of the masker) placed at the time/frequency coordinate of the masker under consideration in the speech landscape of Figure 12.4. The surface of Figure 12.3 will be below the surface of the speech landscape of Figure

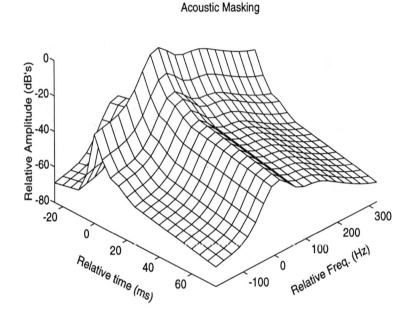

FIGURE 12.3
Psycho-acoustic masking data, both temporal and frequency.

12.4 at some places, and above at others. Figure 12.3 represents the threshold of detectability. When the surface of the speech landscape is below, those sounds cannot be heard. When the surface of the speech landscape is above, those sounds can be heard, relative to the masker under consideration.

This process is repeated for all time/frequency coordinates of the speech landscape, with the appropriate masking surfaces, to determine which sounds are masked by which others.

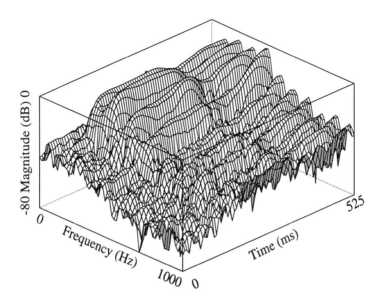

FIGURE 12.4
Time/frequency/magnitude representation of a segment of speech.

12.2 Perceptual Coding Considerations

The discussion of the previous section describes in a conceptual manner the application of simultaneous and temporal masking data to determine which signal components are not perceptible. Two other specific considerations impact practical application of masking in perceptual coding. Standard frequency domain transformations include limits on their time/frequency resolution (size of the x−y grid spacing on the speech landscape). Additionally, masking data sets (the surface of Figure 12.3) differ, depending on whether the masker is tone-like or noise-like, and whether the maskee is tone-like or noise-like.

12.2.1 Limits on Time/Frequency Resolution

Wideband perceptual coding (bandwidth of approximately 20 kHz) is used in audio coding in standards such as MPEG-2 and MPEG-4. The duration of the analysis window used in these wideband coding techniques is around 10 ms [82, 12, 83]. The reciprocal relation between time and frequency dictates that a frequency resolution of only 100 Hz can be obtained using standard frequency-domain transforms (discrete Fourier transform) for this time resolution (10 ms). Here, the frequency resolution refers to the frequency spacing between samples of the DFT. Given this frequency resolution, it is possible to locate regions of simultaneous masking in frequency at high frequencies (i.e., above 5kHz) using the type of data in Figure 6.4. Significant economies in coding can be achieved with frequency domain analysis for wideband audio signals. However, for the lower frequency regions of signals, a much higher frequency resolution is required to exploit the masking properties of the human auditory system. This results from the much narrower frequency spacing of the critical bands at low frequencies.

For temporal masking analysis, a time length as short as 10 ms is useful to take advantage of the qualities of both forward and backward masking. This time resolution is crucial because of the rapid drop off of the amount of masking with time (see Figure 6.5). Longer analysis windows would blend together separate sounds. However, as described, this frequency resolution (100 Hz) is not sufficient to separate the low frequency critical bands for simultaneous masking. To fully exploit the properties of both simultaneous masking and temporal masking, it is necessary to bypass the constraints imposed by the reciprocal relation between time and frequency. Section 12.2.3 suggests a method to circumvent these limitations by utilizing information about the human vocal system.

12.2.2 Sound Quality of Signal Components

Psycho-acoustic experimentation on the auditory system [45, 141, 139, 176, 31] has revealed that a tone masked by a broad band of noise is different from a broad band of noise masked by a broad band of noise, a tone masked by a tone, or a broad band of noise masked by a tone. (It has been shown that a narrow band of noise has similar masking properties as a pure tone [59, 60, 61].) This is because the notches in the plot of Figure 6.4 occur only during tone-on-tone and broadband-noise-on-tone masking. These notches are not present for tone-on-broadband-noise or broadband-noise-on-broadband-noise masking [81, 60]. "The notch has been shown to be caused by the detection on the lower frequency side of the masker, of the combination tones that are produced by the addition of the masker and the signal"[60].

For perceptual coding, it is important to know the characteristics of the signal. With simple signal processing techniques, the speech spectrum of the short-time analysis window is segmented into discrete frequency bands. Each subband is classified as either noise-like or tone-like so it can be determined which type of masking occurs on each subband of the signal.

12.2.3 MBE Model for Perceptual Coding

The Multi Band Excitation (MBE) speech model, discussed in Section 11.1, provides an approach to handle the considerations of the two previous sections: tone-like or noise-like signal component classification and limits on time/frequency resolution.

The MBE model divides the frequency spectrum into frequency bins centered at harmonics of the pitch frequency of the speech signal. The analysis classifies each frequency bin as voiced or unvoiced. Voiced bins are characterized by a pitch harmonic (tone) located at that frequency. Unvoiced bins are characterized by a band of white noise across the frequency bin. This provides an inherent tone-like versus noise-like classification of signal components in the MBE analysis. This classification can be used to select the appropriate masking data for perceptual coding, based on the sound qualities of the masker and maskee.

By assuming that speech follows the basic properties of the MBE speech model, the complex magnitudes of the speech spectra at harmonics of the pitch frequency, and the associated voiced/unvoiced decisions, determine the speech spectra completely. Based on the MBE speech analysis, the temporal resolution of the signal corresponds to the frame rate. Considering psycho-acoustic frequency and temporal masking data and critical bands, a 10 ms temporal resolution and 25 Hz frequency resolution are required to sufficiently determine the masked regions of the signal. The 10 ms temporal resolution is required in order to utilize the strongest aspects of temporal masking, when the maskee and masker are close in time (see Figure 6.5). Because the critical bands in frequency region below 800 Hz are less than 75 Hz, a 25 Hz frequency resolution is needed to ensure at least two frequency bins in most critical bands.

MBE Analysis/Synthesis with Masking

The masking approach described in Section 12.1.3 was applied to speech data analyzed and synthesized using the MBE model without quantization [56, 57]. The voiced/unvoiced classification of frequency bins was used to select appropriate masking data.

Degradation mean opinion score (DMOS) (see Section 8.2.2) listening tests were performed to rate the relative quality of two processing schemes. The first scheme analyzed and synthesized the speech data using the MBE model without quantization or other altering of the model parameters. This data was used as the reference. The second processing algorithm eliminated specific spectral magnitude information as guided by the masking thresholds.

The perceptual processing was designed to yield perceptual quality measurements equal to those of the unaltered MBE parameters. This indicates that the additional auditory processing is functioning transparently, by not adding perceptible degradations.

Figure 12.5 displays the results of the listening tests. A DMOS score of 4 was obtained when the threshold was held at 10dB. A score of 4 indicates no degradation. Although the speech is no longer coded transparently, it is interesting to note that as the threshold level is raised, less perceptually significant information is removed. This is a good technique to use to lower the bit rate of the coder.

12.3   Research in Perceptual Speech Coding

Researchers are actively investigating the field of perceptual speech coding. Current efforts are being directed at two primary concepts: transforming the speech signal into a perceptual representation and distributing quantization noise below masking thresholds. The two are related, and both are necessary to improve coder quality through perceptual considerations.

Johnston [82, 12] was the first to use masking criteria to distribute quantization bits in wideband audio coding. He calculated the perceptual significance of each frequency band in an audio signal using simultaneous masking calculations and distributed quantization bits accordingly. Huang [76] extended these techniques to include forward masking criteria. Johnston's work is now being incorporated into speech coders.

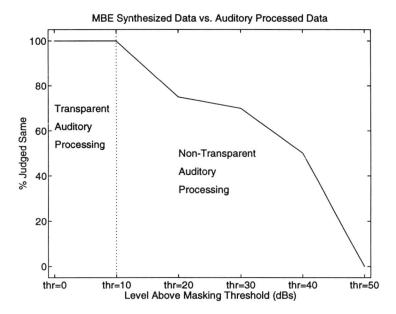

FIGURE 12.5
MBE synthesized data compared to auditory processed data.
Quality degrades when masking threshold is raised above 10
dB. Below 10 dB there is no audible degradation.

Bourget et al. [11], Sen and Holmes [145], and Soheili et al. [149] use perceptual measures in creating the excitation codebook in CELP-based coders. Carnero and Drygajlo [32, 18] decompose the signal into critical bands and use masking thresholding to determine bit allocation. Najafzadeh-Azghandi and Kabal [118, 119] as well as George and Smith [49] use perceptual masking thresholds to train the vector quantization in a sinusoidal based coder.

Both Virag [161] and Drygajlo and Carnero [32] are utilizing acoustic masking techniques for speech enhancement. Kubin and Kleijn are working on computationally efficient perceptual speech coding algorithms [99]. Tang et al. are using perceptual techniques within a subband speech coder [157].

Much work has been directed at using perceptual criteria to distribute coding error to minimize perceived degradation [82, 76, 102, 11, 18, 99, 149, 145, 118, 119]. In [145], Sen and Holmes attempt to shape the error spectrum of a CELP coder such that it is below the calculated masking threshold. This method quantizes the areas of the speech spectrum which lie above the masking threshold with minimal distortion, while quantizing those regions below the masking threshold with as much distortion as the masking threshold will allow. Reducing the coding bit rate and, correspondingly, raising the quantization noise above the masking threshold introduced only minor perceptual distortion. The reported perceived effect of this coder is much smoother, more natural sounding decoded speech than typical CELP encoded/decoded speech at the same bit rate.

Drygajlo and Carnero approach coding and speech quality enhancement in the same algorithm [32]. The method uses wavelet decomposition (a transformation, similar to an FFT, but with unevenly spaced frequency basis functions placed to resemble critical bands) to obtain frequency responses of critical bands to help efficiently calculate masking thresholds. Coding bits are allocated such that the quantization noise remains below the masked threshold of detectability.

Appendix A

Related Internet Sites

The Internet supplies a wealth of information on speech coding. Specifications of coding standards can be downloaded, along with listings of conference proceedings, and even example software for various coders.

However, because of the dynamic nature of the Internet, the pointers listed below can expire at any time. If such is the case, backing up higher in the directory structure (possibly to the home page of the site) and following likely branches might lead to the same information, arranged differently. Most of the pages contained at these sites are protected by copyrights. This listing is, by no means, complete and is presented as a starting point for further information.

A.1 Information on Coding Standards

ITU

The International Telecommunication Union (ITU) is one of the primary standards organizations. Formerly known as the CCITT, the telecommunications section is the ITU-T. The primary page for the ITU is located at:

`www.itu.int`

ITU speech coding standards fall in the "G" series of recommendations, including G.711 (μ law), G.728 (16 kbit/s Low Delay CELP), G.729A (8.0 kbit/s CS-ACELP), etc. Very brief descriptions and tables of content of the speech coding standards are located at:

www.itu.int//itudoc/itu-t/rec/g/g700-799/index.html

Electronic versions of the standards documentation can be purchased at the same location.

ETSI

The European Telecommunications Standards Institute (ETSI) publishes standards for digital cellular communications in Europe. The Special Mobile Group (SMG) 11 is responsible for speech coding algorithms and the corresponding standards including GSM 06.60 Enhanced Full Rate (EFR, 12.2 kbit/s ACELP) and 06.90 Adaptive Multi-Rate (AMR). The main page for ETSI is located at:

www.etsi.org

By selecting the "Publications and Products" page, it is possible to download electronic versions of the standards documentation.

ISO/MPEG

The International Standards Organization (ISO) is the parent organization for the Moving Picture Experts Group (MPEG). MPEG standards are primarily concerned with video and general audio coding. However, with MPEG 4, two speech coders have been standardized to cover a wide range of bit rates. The parametric Harmonic Vector Excitation (HVXC) coder operates at 2 or 4 kit/s. The narrowband CELP encodes speech at 6 to 12 kbits/s. The wideband CELP operates at 18 kbits/s with the speech sampled at 16 kHz.

The MPEG home page is at:

drogo.cselt.stet.it/mpeg

It contains general information about the standards and current workings of the different groups.

Documentation on the parametric and the CELP speech coders are contained in the files:

ftp://ftp.tnt.uni-hannover.de/pub/MPEG/audio/mpeg4/
documents/w2203/w2203par.pdf

ftp://ftp.tnt.uni-hannover.de/pub/MPEG/audio/mpeg4/
documents/w2203/w2203clp.pdf

A report of subjective listening tests is available at:

www.tnt.uni-hannover.de/project/mpeg/audio/public/w2424.html

The report compares the 2.0 and 4.0 kbit/s parametric Harmonic Vector Excitation (HVXC) to the 4.8 kbit/s U.S. Federal Standard 1016 (FS1016). The multiple bit rate (6.0, 8.3, 12.0 kbit/s) CELP algorithm is compared to the ITU-T G.723.1 and G.729 standards and the ETSI GSM-EFR standard.

DDVPC

The U.S. Department of Defense Voice Processing Consortium (DDVPC) standardizes speech coders for U.S. government applications. The Federal Standard 1015 (FS1015) LPC-10e is a two-state 2.4 kbit/s LPC coder. The FS1016 CELP coder compresses speech to 4.8 kbit/s. The new Federal standard 2.4 kbit/ MELP coder is also documented at this location. Example software and example coded/decoded speech is available for all three coders.

The main page for the DDVPC is located at:

www.plh.af.mil/ddvpc/index.html

Pages for the example coded speech and software are referenced from the main page.

A.2 Technical Conferences

ICASSP

The Signal Processing Society of the IEEE sponsors the annual International Conference on Acoustics, Speech, and Signal Processing (ICASSP). ICASSP devotes 4 to 5 technical sessions to speech coding, with each session containing 8 papers. The broad coverage of ICASSP speech coding sessions includes most all topics of current research interest. The IEEE home page is located at:

www.ieee.org

Each year, the host for the conference provides the web site. For 2000, the main page is:

`icassp2000.sdsu.edu`

Information on the conference from 1999, can be found at the main page of:

`icassp99.asu.edu`

while the technical programs for speech, including abstracts, are located at:

`icassp99.asu.edu/technical/sessions/program-SP.html`

Information for the 1997 conference can be found at the main page:

`www.nt.e-technik.uni-erlangen.de/icassp97`

The technical sessions are listed in the "Program" page. The 1998 conference listing, formerly hosted by Microsoft, is no longer available.

IEEE Speech Coding Workshops

The Signal Processing Society of the IEEE organizes the biennial Speech Coding Workshop. The workshop falls on odd years and covers topics of current research in detail. A listing of the paper titles for the 1999 conference is located at:

`sigwww.cs.tut.fi/TICSP/SCW99/program.htm`

References

[1] J. Adoul, P. Mabilleau, M. Delprat, and S. Morisette. Fast CELP coding based on algebraic codes. *IEEE Int. Conf. Acoust. Sp. Sig. Proc.*, 1987, pp. 1957–1960.

[2] N. Ahmed and K. Rao. *Orthogonal Transforms for Digital Signal Processing.* Springer Verlag, New York, 1975.

[3] B. Atal and S. Hanauer. Speech analysis and synthesis by linear prediction of the speech wave. *J. Acoust. Soc. Am.* 1971, pp. 637–655.

[4] B.S. Atal and J.R. Remde. A new model of lpc excitation for producing natural-sounding speech at low bit rates. *IEEE Int. Conf. Acoust. Sp. Sig. Proc.*, 1982, pp. 614–617.

[5] B.S. Atal. High-quality speech at low bit rates:multi-pulse and stochastically excited linear predictive coders. *IEEE Int. Conf. Acoust. Sp. Sig. Proc.*, 1986, pp. 1681–1684.

[6] I. Atkinson, S. Yeldener, and A. Kondoz. High quality split band LPC vocoder operating at low bit rates. *IEEE Int. Conf. Acoust. Sp. Sig. Proc.;* 1997, pp. 1559–1562.

[7] T. Barnwell. Recursive windowing for generating autocorrelation coefficients for LPC analysis. *IEEE Trans. Acoust. Sp. Sig. Proc.*, 1981, pp. 1062–1066.

[8] J. Beerends and J. Stemerdink. A perceptual speech quality measure based on a psychoacoustic sound representation. *J. Audio Eng. Soc.*, 1994, pp. 115–123.

[9] R. Bellman. *Dynamic Programming* Princeton University Press, Princeton, NJ, 1957.

[10] L.L. Beranek. *Acoustics.* American Institute of Physics, New York, 1986.

[11] C. Bourget, T. Aboulnasr, and E. Verreault. Perceptual speech coder. *IEEE Can. Conf. Elec. Comp. Eng.*, 1995, Vol. 2:1070–1072.

[12] K. Brandenburg and J. D. Johnston. Second generation perceptual audio coding: The hybrid coder. *Audio Eng. Soc. Proc. Preprint*, Mar, 1990.

[13] G. Bristow. *Electronic Speech Synthesis.* McGraw Hill, New York, 1984.

[14] S. Campanella et al. A comparison of orthogonal transformations for digital speech processing. *IEEE Trans. Comm.*, 1971, COM-19:1045.

[15] J. Campbell and T. Tremain. Voiced/unvoiced classification of speech with applications to the U.S. Government LPC-10e algorithm. *IEEE Int. Conf. Acoust. Sp. Sig. Proc.*, 1986, pp. 473–476.

[16] J. Campbell, V. Welch, and T. Tremain. An Expandable Error-Protected 4800 BPS CELP Coder (U.S. Federal Standard 4800 BPS Voice Coder). *IEEE Int. Conf. Acoust. Sp. Sig. Proc.*, 1989, pp. 735–738.

[17] J.P. Carlson. Digitalized phase vocoder. *Proc. Conf. Sp. Comm. Proc.*, Nov 1967.

[18] B. Carnero and A. Drygajloa. Perceptual speech coding using time and frequency masking constraints. *IEEE Int. Conf. Acoust. Sp. Sig. Proc.*, 1997, pp. 1363–1366.

[19] J. Chen and A. Gersho. Real-time VAPC speech coding at 4800 bit/s with adaptive postfiltering. *IEEE Int. Conf. Acoust. Sp. Sig. Proc.*, 1987, pp. 2185–2188.

[20] J. Chen. High quality 16 kbit/s speech coding with a one-way delay less than 2 ms. *IEEE Int. Conf. Acoust. Sp. Sig. Proc.*, 1990, pp. 453–456.

[21] J. Collura and T. Tremain. Vector quantizer design for the coding of LSF parameters. *IEEE Int. Conf. Acoust. Sp. Sig. Proc.*, 1992, pp. II–29 – II–32.

[22] J.B. Costello and F.S. Mozer. Time domain synthesis gives good quality speech at very low data rates. *Sp. Tech.*, 1983, Vol. 1, No.3:62–68.

[23] R.E. Crochiere and J.L. Flanagan. The technology of digital speech compression, editing and storage. *Nat. Comp. Conf. Rec.*, May 1983.

[24] R.E. Crochiere, S.A. Webber, and J.L. Flanagan. Digital coding of speech in subbands. *Bell Syst. Tech. J.*, 1976, 55:1069–1085.

[25] P. Cummiskey, N.S. Jayant, and J.L. Flanagan. Adaptive quantization in differential pcm coding of speech. *Bell Sys. Tech. J.*, Sep 1973, 52-7:1105–1118.

[26] V. Cuperman and A. Gersho. Adaptive differential vector coding of speech. *Proc. Globecom*, Dec 1982, 52-7:E6.6.1.

[27] V. Cuperman and A. Gersho. Vector predictive coding of speech at 16 kb/s. *IEEE Trans. on Communications*, Jul 1985, COM-33:685.

[28] B.H. Deatherage and T.R. Evans. Binaural masking: Backward, forward, and simultaneous effects. *J. Acoust Soc. Am.*, 1969, pp. 362–371.

[29] P.B. Denes and E.N. Pinson. *The Speech Chain: The Physics and Biology of Spoken Language.* Waverly Press, Baltimore, 1963.

[30] S. Dimolitsas. Evaluation of voice codec performance for the Inmarsat Mini-M system. *Tenth Int. Conf. Digital Sat. Comm.*, 1995.

[31] D.D. Dirks and D. Bower. Effect of forward and backward masking on speech intelligibility. *J. Acoust Soc. Am.*, 1970, pp. 1003–1008.

[32] A. Drygajlo and B. Carnero. Integrated speech enhancement and coding in the time-frequency domain. *Proc. IEEE Int. Conf. Acoust. Sp. Sig. Proc.*, 1997, pp. 1183–1186.

[33] H. Dudley. The vocoder. *Bell Labs Rec.*, 1939, 17:122–126.

[34] L.L. Elliot. Backward masking: Monotic and dichotic conditions. *J. Acoust Soc. Am.*, Aug 1962, pp. 1108–1115.

[35] D. Elliott and K. Rao. *Fast Transforms – Algorithms*. Academic Press, New York, 1982.

[36] G. Fairbanks. Test of phonemic variation: The rhyme test. *J. Acoust. Soc. Am.*, 1958, pp. 596–600.

[37] G. Fant. *Acoustic Theory of Speech Production*. Mouton and Co. N.V., the Hague, the Netherlands, 1970.

[38] J. L. Flanagan. *Speech Analysis, Synthesis and Perception*. Springer-Verlag, New York, 1972.

[39] J.L. Flanagan, M.R. Scroeder, B.S. Atal, R.E. Crochiere, N.S. Jayant, and J.M. Tribolet. Speech coding. *IEEE Trans. Comm.*, 1979, COMM-27:710–737.

[40] J.L. Flanagan. Parametric coding of speech spectra. *J. Acoust. Soc. Am.*, Aug 1980, JASA-68:412–419.

[41] J.L. Flanagan. *Speech Analysis, Synthesis, and Perception*. Springer-Verlag, New York, 1965.

[42] J.L. Flanagan. *Speech Analysis, Synthesis, and Perception*. Springer-Verlag, New York, 1972.

[43] J.L. Flanagan and B. J. Watson. Binaural unmasking of complex signals. *J.. Acoust. Soc. Am.*, Aug 1966, 40:456–468.

[44] J.L. Flanagan, K. Ishizaka, and K.L. Shipley. Signal models for low bit-rate coding of speech. *J. Acoust Soc. Am.*, Sep 1980, JASA-68:780–791.

[45] H. Fletcher. *Speech and Hearing*. D. Van Nostrand, New York, 1929.

[46] H. Fletcher. *Speech and Hearing in Communication*. D. Van Nostrand, New York, 1953.

[47] J. Foster and R. Gray. Finite state vector quantization for waveform coding. *IEEE Trans.*, May 1985, IT-31:348.

[48] O. Fujimara. An approximation to voice aperiodicity. *IEEE Trans. Audio Electroacoust.*, Mar 1968, pp. 68–72.

[49] E.B. George and M.J. Smith. Perceptual considerations in a low bit rate sinusoidal vocoder. *IEEE Int. Conf. Comp. Comm.*, 1990, pp. 268–275.

[50] J. Gibson. Adaptive prediction for speech encoding. *ASSP Magazine*, 1984, 1:12–26.

[51] B. Gold. Computer program for pitch extraction. *J. Acoust. Soc. Am.*, 1962, JASA-37:753–754.

[52] B. Gold and C. Rader. The channel vocoder. *IEEE Trans. Audio Electroacoust.*, Dec 1967, pp. 148–160.

[53] B. Gold and C. Rader. Systems for compressing the bandwidth of speech. *IEEE Trans. Audio Electroacoust.*, Sep 1967, pp. 131–135.

[54] B. Gold and J. Tierney. Vocoder analysis based on properties of the human auditory system. *M.I.T. Lincoln Laboratory Tech. Rep.*, TR-670, Dec 1983.

[55] B. Gold and L.R. Rabiner. Parallel processing techniques for estimating pitch periods of speech signals in the time domain. *J. Acoust. Soc. Am.*, Aug 1969, pp. 442–448.

[56] R.G. Goldberg. Perceptual Speech Coding. PhD Dissertation: Rutgers University, New Brunswick, NJ, Nov 1993.

[57] R.G. Goldberg and J. L. Flanagan. Perceptual Speech Coder and Method. *US Patent US5706392*, Jun 1995.

[58] O. Gottesman and A. Gersho Enhanced waveform interpolative coding at 4 kbps. *IEEE Speech Coding Workshop*, 1999.

[59] D.D. Greenwood and J.M. Goldberg. Response of neurons int the cochlear nuchlei to variations in noise bandwidth and to tone-noise combinations. *J. Acoust. Soc. Am.*, Sep 1970, JASA-47:1022–1040.

[60] D.D. Greenwood. Masking, combination tones, and critical bandwidth. *J. Acoust. Soc. Am.*, Jan 1970, JASA-47-1:108.

[61] D.D. Greenwood. Aural combination tones and auditory masking. *J. Acoust. Soc. Am.*, Aug 1971, JASA-50:502–543.

[62] D.W. Griffin. The multi-band excitation vocoder. PhD Dissertation: MIT, Cambridge, MA, Feb 1987.

[63] D.W. Griffin and J.S. Lim. A new model-based speech analysis/synthesis system. *IEEE Int. Conf. Acoust. Sp. Sig. Proc.*, Mar 1985, pp. 513–516.

[64] D.W. Griffin and J.S. Lim. A high quality 9.6 kbps speech coding system. *IEEE Int. Conf. Acoust. Sp. Sig. Proc.*, 1986.

[65] D.W. Griffin and J.S. Lim. Multiband excitation vocoder. *IEEE Trans. Acoust. Sp. Sig. Proc.*, Aug 1988, Vol. 36 No. 8, pp. 1223-1235.

[66] J. Hardwick and J. Lim. A 4800 bps improved multi-band excitation speech coder. *IEEE Speech Coding Workshop*, 1989.

[67] J. Hardwick and J. Lim. Application of the IMBE speech coder to mobile communications. *IEEE Int. Conf. Acoust. Sp. Sig. Proc.*, 1991.

[68] S. Heinen, M. Adrat, O. Vary, and W. Xu. A 6.1 to 13.3 kb/s variable rate CELP codec for AMR speech coding. *IEEE Int. Conf. Acoust. Sp. Sig. Proc.*, 1999, pp. 9-12.

[69] H. Hermansky. Perceptual linear predictive (PLP) analysis of speech. *J. Acoust Soc. Am.*, 1990, JASA-87:1738-1752.

[70] W. Hess. *Pitch Determination of Speech Signals*. Springer-Verlag, New York, 1983.

[71] J.E. Hind. Two-tone masking effects in squirrel monkey auditory nerve fibers. *Freq. Anal. Period. Detect. Hear.*, 1970.

[72] I.J. Hirsh. Auditory perception of temporal order. *J. Acoust Soc. Am.*, Jun 1959, JASA-31:759-767.

[73] T. Hokanen, J. Vainio, K. Jarvinen, P. Haavisto, R. Salami, C. Laflamme, and J. Adoul. Enhanced full rate speech codec for IS-136 digital cellular system. *IEEE Trans. Acoust. Sp. Sig. Proc.*, 1997, pp. 731-734.

[74] J.N. Holmes. The JSRU channel vocoder. *IEE Proc.*, Feb 1980, 127-1:53-60.

[75] A. House, C. Williams, M. Hecker, and K. Kryter. Articulation testing methods: consonantal differentiation with a closed-response set. *J. Acoust. Soc. Am.* 1965, pp. 158-166.

[76] Y. Huang and T. Chiueh. A new forward masking model and its application to perceptual audio coding. *IEEE Int. Conf. Acoust. Sp. Sig. Proc.*, 1999, pp. 905-908.

[77] K. Jarvinen, J. Vainio, P. Kapanen, T. Hokanen, P. Haavisto, R. Salami, C. Laflamme, and J. Adoul. GSM enhanced full rate speech codec. *IEEE Trans. Acoust. Sp. Sig. Proc.*, 1997, pp. 771-774.

[78] N.S. Jayant. Adaptive quantization with a one-word memory. *The Bell Sys. Tech. J.*, Sep 1973, 52-7:1119–1144.

[79] N.S. Jayant and P. Noll. *Digital Coding of Waveforms:Principles and Applications to Speech and Video*. Prentice Hall, Englewood Cliffs, NJ, 1984.

[80] N.S. Jayant, V.B. Lawrence, and D.P. Prezas. Coding of speech and wideband audio. *ATT Tech. J.*, Sep/Oct 1990, pp. 25–41.

[81] L. A. Jeffress. Masking. Chapter 3 in *Foundation of Modern Auditory Theory, Vol. I*. Edited by J.V. Tobias. Academic Press, New York, 1970.

[82] J.D. Johnston. Transform coding of audio signals using perceptual noise criteria. *IEEE J. Select. Areas Comm.*, 1988, 6:314–322.

[83] J. D. Johnston. Perceptual transform coding of wideband stereo signals. *IEEE Int. Conf. Acoust. Sp. Sig. Proc.*, 1989, pp. 1993–1996.

[84] J.D. Johnston. Transform coding of audio signals using perceptual noise criteria. *IEEE J. on Select. Areas Comm.*, Feb 1989, pp. 314–323.

[85] J.D. Johnston and K.H. Brandenburg. Sound coding algorithm. *MPEG-891-148*, Report of ISO-IEC/JTCI/SC2/WG8, 1989.

[86] P. Kabel and R. Ramachandran. The computation of line spectral frequencies using Chebyshev polynomials. *IEEE Trans. Acoust. Sp. Sig. Proc.*, 1986, pp. 1419–1426.

[87] H. Kang and D. Sen. Phase adjustment in waveform interpolation. *IEEE Int. Conf. Acoust. Sp. Sig. Proc.*, 1999, pp. 261–264.

[88] G.S. Kang and S.S. Everet. Improvement of the excitation source in the narrow-band linear prediction vocoder. *IEEE Trans. Acoust. Sp. Sig. Proc.*, Apr 1985, ASSP-33:317–386.

[89] W.D. Keidel and W.D. Neff. Adaptation and masking. Chapter 8 of *Handbook of Sensory Physiology*, 1976, pp. 689–705.

[90] W.B. Kleijn et al. Improved speech quality and efficient vector quantization in selp. *IEEE Int. Conf. Acoust. Sp. Sig. Proc.*, 1988, pp. 155–158.

[91] W. Kleijn, D. Krasinski, and R. Ketchum. Fast methods for the CELP speech coding algorithm. *IEEE Trans. Acoust. Sp. Sig. Proc.*, 1990, pp. 1330–1341.

[92] W. Kleijn. Encoding speech using prototype waveforms. *IEEE Trans. Acoust. Sp. Audio Proc.*, 1993, pp. 386–399.

[93] W. Kleijn and J. Haagen. A speech coder based on decomposition of characteristic waveforms. *IEEE Int. Conf. Acoust. Sp. Sig. Proc.*, 1995, pp. 508–511.

[94] W. Kleijn, Y. Shoham, D. Sen, and R. Hagen. A low-complexity waveform interpolation coder. *IEEE Int. Conf. Acoust. Sp. Sig. Proc.*, 1996, pp. 212–215.

[95] M. Kohler. A comparison of the new 2400 bps MELP Federal Standard with other standard coders. *IEEE Int. Conf. Acoust. Sp. Sig. Proc.*, 1997, pp. 1587–1590.

[96] A. Kondoz. *Digital Speech: Coding for Low Bit Rate Applications* John Wiley & Sons, Chichester, U.K., 1994.

[97] P. Kroon, E.F. Deprettere, and R.J. Sluyter. Regular-pulse excitation – a novel approach to effective and efficient multipulse coding of speech. *IEEE Trans. Acoust. Sp. Sig. Proc.*, Oct 1986, ASSP-34:1054–1063.

[98] P. Kroon and B.S. Atal. Pitch predictors with high temporal resolution. *IEEE Int. Conf. Acoust. Sp. Sig. Proc.*, Apr 1990, ICASSP-90:661–664.

[99] G. Kubin and W. Kleijn. On speech coding in a perceptual domain. *Proc. IEEE Int. Conf. Acoust. Sp. Sig. Proc.*, 1999, pp. 205–208.

[100] S.Y. Kwon and A.J. Goldberg. An enhanced LPC vocoder with no voiced/unvoiced switch. *IEEE Trans. Acoust. Sp. Sig. Proc.*, Aug 1984, ASSP-32:851–858.

[101] P. LeBlanc, B. Bhattacharya, S Mahmoud, and V. Cuperman. Efficient search and design procedures for robust multi-stage VQ of LPC parameters 4 kb/s speech coding. *IEEE Trans. Sp. Audio Proc.*, Oct 1993.

[102] E. Levine. Stochastic Vector Quantization Using Neural Networks. PhD Dissertation, Stanford University, Stanford, CA, 1996.

[103] Y. Linde, A. Buzo, and R. Gray An algorithm for vector quantizer design. *IEEE Trans. Comm.*, 1980, pp. 84–95.

[104] J. Makhoul, R. Viswanathan, R. Schwartz, and A.W.F. Hugins. A mixed-source excitation model for speech compression and synthesis. *IEEE Int. Conf. Acoust. Sp. Sig. Proc.*, Apr 1978, pp. 163–166.

[105] J. Makhoul. Linear Prediction: A tutorial review. *Proc. IEEE*, 1975, pp. 561–580.

[106] J. Makhoul et. al. A mixed-source model for speech compression and synthesis. *J. Acoust. Soc. Am.*, 1978, pp. 1577–1581.

[107] J. Makhoul, S. Roucos, and H. Gish. Vector quantization for speech coding. *Proc. IEEE*, 1985, pp. 1551–1588.

[108] M. Marcellin, T. Fisher, and J. Gibson. Predictive trellis coded quantization of speech. *IEEE Trans. Sp. Sig. Proc.*, Jan 1990, ASSP-38:46.

[109] J. Markel and A. Gray. *Linear Prediction of Speech.* Springer Verlag, Berlin, 1976.

[110] R.J. McAuley and T.F. Quatieri. Sine-wave phase coding at low data rates. *IEEE Int. Conf. Acoust. Sp. Sig. Proc.*, 1991, pp. 577–580.

[111] R.J. McAuley and T.F. Quatieri. The application of subband coding to improve quality and robustness of the sinusiodal transform coder. *IEEE Int. Conf. Acoust. Sp. Sig. Proc.*, 1991, pp. 577–580.

[112] R.J. McAuley and T.F. Quatieri. Speech analysis/synthesis based on a sinusoidal representation. *IEEE Trans. Acoust. Sp. Sig. Proc.*, 1986, ASSP-34:744–754.

[113] R.J. McAuley, T.F. Quatieri. Chapter 6 in *Advances In Speech Signal Proccessing.* Edited by S. Furui, and M.M. Sondi. Marcel Dekker, New York, 1992.

[114] A. McCree and T. Barnwell. A mixed excitation LPC vocoder model for low bit rate speech coding. *IEEE Trans. on Speech and Audio Processing*, 1995, pp. 242–250.

[115] A. McCree, K. Truong, E. George, T. Barnwell, and V. Viswanathan. A 2.4 kbit/s MELP coder candidate for the new U. S. Federal standard. *IEEE Int. Conf. Acoust. Sp. Sig. Proc.*, 1996, pp. 200–203.

[116] A. McCree and J. De Martin. A 1.7 kb/s MELP coder with improved analysis and quantization. *IEEE Int. Conf. Acoust. Sp. Sig. Proc.*, 1998, pp. 593–596.

[117] W. Mikhael and A. Spanias. Accurate representation of time-varying signals using mixed transforms with applications to speech. *IEEE Trans.*, Feb 1989, CAS-36, No. 2:329.

[118] H. Najafzadeh-Azghandi and P. Kabal. Perceptual coding of narrowband audio signals at 8 kbits/s. *Sp. Coding for Telecomm. Proc.*, 1997, pp. 109–110.

[119] H. Najafzadeh-Azghandi and P. Kabal. Improving perceptual coding of narrowband audio signals at low rates. *IEEE Int. Conf. Acoust. Sp. Sig. Proc.*, 1997, pp. 109–110.

[120] H. Ney. A dynamic programming technique for nonlinear smoothing. *IEEE Int. Conf. Acoust. Sp. Sig. Proc.*, 1981, pp. 62–65.

[121] M. Nishiguchi, A. Inoue, Y. Maeda, and J. Matsumoto Parametric speech coding HVXC at 2.0–4.0 kbps. *Proc. IEEE Speech Coding Workshop*, 1999.

[122] A.M. Noll. Cepstrum pitch determination. *J. Acoust. Soc. Am.*, Feb 1967, JASA-41:293–309.

[123] P. Noll. Non-adaptive and adaptive dpcm of speech signals. *Polytech Tijdschr. Ed. Elektrotech./Electron*, the Netherlands, No. 19, 1972.

[124] B. Novorita. Incorporation of temporal masking effects into bark spectral distortion measure. *IEEE Int. Conf. Acoust. Sp. Sig. Proc.*, 1999, pp. 665–668.

[125] A. Oppenheim and R. Schafer. *Digital Signal Processing.* Prentice Hall, Englewood Cliffs, NJ, 1975.

[126] A. Oppenheim and R. Schafer. *Discrete Time Signal Processing.* Prentice Hall, Englewood Cliffs, NJ, 1989.

[127] S.J. Orfanidis. *Introduction to Signal Processing.* Prentice Hall, Upper Saddle River, NJ, 1995.

[128] D. O'Shaughnessy. *Speech Communication: Human and Machine.* Addison-Wesley, Reading, MA, 1987.

[129] K. Paliwal and B. Atal. Efficient vector quantization of LPC parameters at 24 bits/frame. *IEEE Int. Conf. Acoust. Sp. Sig. Proc.*, 1991, pp. 661–664.

[130] K. Paliwal and B. Atal. Efficient vector quantization of LPC parameters at 24 bits/frame. *IEEE Trans. Sp. Audio Proc.*, 1993, pp. 3–14.

[131] P.E. Papamichalis. *Practical Approaches to Speech Coding.* Prentice-Hall, Englewood Cliffs, NJ, 1987.

[132] D. Paul. A 500–800 b/s adaptive vector quantization vocoder using a perceptually motivated distance measure. *Proc. Globecom*, Dec 1982, ASSP-38:E6.3.1.

[133] E. Peterson and F.S. Cooper. Peakpicker: A bandwidth compression device. *J. Acoust. Soc. Am.*, Jun 1957, JASA-29:777–782.

[134] J.M. Pickett. Backward masking. *J. Acoust Soc. Am.*, Dec 1959, JASA-31:1613–1615.

[135] M.R. Portnoff. A quasi-one-dimensional digital simulation for the time-varying vocal tract. M.S. Thesis: MIT, Cambridge, MA, Jun 1973.

[136] L. Rabiner, L. Sambur, and C. Schmidt. Applications of nonlinear smoothing algorithm to speech processing. *IEEE Trans. Acoust. Sp. Sig. Proc.*, 1975, pp. 552–557.

[137] L.R. Rabiner and R.W. Schafer. *Digital Processing of Speech Signals.* Prentice-Hall, Englewood Cliffs, NJ, 1978.

[138] E. Riskin and R. Gray. A greedy tree growing algorithm for the design of variable rate vector quantizers. *IEEE Trans. on Sig. Proc.*, Nov 1991.

[139] M.B. Sachs. Ed. *Physiology of the Auditory System: A Workshop.* National Educational Consultants, Baltimore, 1971.

[140] I.K. Samoilova. Masking of short tone signals as a function of the time interval between masked and masking sounds. *J. of Biophys.*, Jan 1959, 4:44–52.

[141] B. Scharf Critical Bands. Chapter 5 in *Foundation of Modern Auditory Theory, Volume I.* Edited by J.V. Tobias. Academic Press, New York, 1970.

[142] M. Schroeder and B. Atal. Code-excited linear prediction (CELP): high quality speech at very low bit rates. *IEEE Int. Conf. Acoust. Sp. Sig. Proc.*, 1985, pp. 937–940.

[143] M.R. Schroeder. Vocoders: Analysis and synthesis of speech. *Proc. of the IEEE*, May 1966, 54-5:720–734.

[144] M.R. Schroeder. Reference signal for signal quality studies. *J. Acoust. Soc. Am.*, Oct 1968, 44-6:1735–1736.

[145] D. Sen and W.H. Holmes. Perceptual enhancement of CELP speech coders. *IEEE Int. Conf. Acoust. Sp. Sig. Proc.*, 1994, pp. 105–108.

[146] Y. Shoham. High-quality speech coding at 2.4 to 4.0 kbps based on time frequency interpolation. *IEEE Int. Conf. Acoust. Sp. Sig. Proc.*, 1993, pp. II-167 – II-170.

[147] S. Singhal and B. Atal Improving the performance of multipulse coders at low bit rates. *IEEE Int. Conf. Acoust. Sp. Sig. Proc.*, 1984.

[148] S. Singhal and B. Atal. Amplitude optimization and pitch prediction in multipulse coders. *IEEE Trans. Acoust. Sp. Sig. Proc.*, 1989, pp. 317–327.

[149] R. Soheili, A.M. Kondoz, and B.G. Evans. An 8 kb/s LC-CELP with improved excitation and perceptual modelling. *IEEE Int. Conf. Acoust. Sp. Sig. Proc.*, 1993, pp. 616–619.

[150] M.M. Sondhi. New method of pitch extraction. *IEEE Trans. Audio Electroacoust.*, Jun 1968, AU16-2:262–266.

[151] F. Soong and B. Huang. Line spectrum pairs (LSP) and speech data compression. *IEEE Int. Conf. Acoust. Sp. Sig. Proc.*, 1984, pp. 1.10.1–1.10.4.

[152] A. Spanias. A hybrid transform method for analysis/synthesis of speech. *Sig. Proc. Magazine*, Aug 1991, pp. 217–229.

[153] A. Spanias and P. Loizou. A mixed fourier/walsh transform scheme for speech coding at 4 kbits/s. *Proc IEE*, Oct 1992, Part I:473–481.

[154] J. Stachurski, A. McCree, and V. Viswanathan. High quality MELP coding at bit-rates around 4 kb/s. *IEEE Int. Conf. Acoust. Sp. Sig. Proc.*, 1999, pp. 485–488.

[155] G.A. Studebaker. *Modern Developments in Audiology – Auditory Masking.* Edited by J. Jerger. Academic Press, New York, 1973.

[156] L. Supplee, R. Cohn, and J. Collura. MELP: The new Federal standard at 2400 bps. *IEEE Int. Conf. Acoust. Sp. Sig. Proc.*, 1997, pp. 1591–1594.

[157] B. Tang, A. Shen, A. Alwan, and G. Pottie Perceptually-based embedded subband speech coder. *IEEE Trans. Sp. Audio Proc.*, Mar 1997.

[158] J. Tardelli and E. Kreamer. Vocoder intelligibility and quality test methods. *IEEE Int. Conf. Acoust. Sp. Sig. Proc.*, 1996, pp. 1145–1148.

[159] J. Tribolet and R. Crochiere. Frequency domain coding of speech. *IEEE Trans. Acoust. Sp. Signal Proc.*, Oct 1979, ASSP-27:512.

[160] T. Unno, T. Barnwell, and K. Truong. An improved mixed excitation linear prediction (MELP) coder. *IEEE Int. Conf. Acoust. Sp. Sig. Proc.*, 1999, pp. 245–248.

[161] N. Virag. Speech enhancement based on masking perperties of the auditory system *Proc. IEEE Int. Conf. Acoust. Sp. Sig. Proc.*, 1995, pp. 796–799.

[162] W. Voiers. Diagnostic acceptability measure for speech communication systems. *IEEE Int. Conf. Acoust. Sp. Sig. Proc.*, 1977, pp. 204–207.

[163] W. Voiers. Evaluating processed speech using the diagnostic rhyme test. *Sp. Tech.*, Jan 1983.

[164] G.Von Békésy. Shearing microphones produced by vibrations near the inner and outer hair cells. *J. Acoust. Soc. Am.*, 1953, pp. 786–790.

[165] G.Von Békésy. *Experiments in hearing.* McGraw Hill, New York, 1960.

[166] H. Von Helmholtz. *On the Sensations of Tone.* Dover Publications, New York, 1954.

[167] S. Wang, A. Sekey, and A. Gersho. An objective measure for predicting subjective quality of speech coders. *IEEE J. Select. Areas in Comm.*, 1992, pp. 819–829.

[168] S.W. Wong. An evaluation of 6.4kbit/s speech codecs for Inmarsat-M system. *IEEE Int. Conf. Acoust. Sp. Sig. Proc.*, 1991.

[169] W. Yang, M. Benbouchta, and R. Yantorno. Performance of the modified bark spectral distortion measure as an objective speech quality measure. *IEEE Int. Conf. Acoust. Sp. Sig. Proc.*, 1998, pp. 541–544.

[170] W. Yang and R. Yantorno. Improvement of MBSD by scaling noise masking threshold and correlation analysis with MOS difference instead of MOS. *IEEE Int. Conf. Acoust. Sp. Sig. Proc.*, 1999, pp. 673–676.

[171] S. Yeldener, A. Kondoz, and B. Evans. A high quality speech coding algorithm suitable for future INMARSAT systems. *Proc. of 7th Euro. Sig. Proc. Conf.*, 1994, pp. 407–410.

[172] S. Yeldener. A 4 kb/s toll quality harmonic excitation linear predictive speech coder. *IEEE Int. Conf. Acoust. Sp. Sig. Proc.*, 1999, pp. 481–484.

[173] M. Yong, G. Davidson, and A. Gersho. Encoding of LPC spectral parameters using switched adaptive interframe prediction. *IEEE Int. Conf. Acoust. Sp. Sig. Proc.*, 1988, pp. 402–405.

[174] H. Zarrinkoub and P. Mermelstein. Switched prediction and quantization of LSP frequencies. *IEEE Int. Conf. Acoust. Sp. Sig. Proc.*, 1995, pp. 757–760.

[175] R. Zelinski and P. Noll. Approaches to adaptive transform coding at low bit rates. *IEEE Trans. Acoust. Sp. and Sig. Proc.*, Feb 1979, ASSP-27:89.

[176] E. Zwicker. Temporal effects in psychoacoustical excitation. *Basic Mech. in Hearing*, 1973, pp. 809–825.

[177] J.J Zwislocki. Central Masking and Auditory Frequency Selectivity. Chapter 3 in *Frequency Analysis and Periodicity in Hearing*. Edited by R. Plomp and G.F. Smoorenburg. A. W. Sijhoff, Leiden, 1970.

[178] European Telecommunications Standards Institute. GSM Adaptive Multi Rate Speech Transcoding (GSM 06.90). ETSI standards documentation, EN 301 704, 1999.

[179] European Telecommunications Standards Institute. GSM Enhanced Full Rate Speech Transcoding (GSM 06.60). ETSI standards documentation, EN 301 245, 1998.

[180] European Telecommunications Standards Institute. GSM Full Rate Speech Transcoding (GSM 06.10). ETSI standards documentation, EN 300 961, 1995.

[181] European Telecommunications Standards Institute. GSM Half Rate Speech Transcoding (GSM 06.20). ETSI standards documentation, EN 300 969, 1998.

[182] Federal Standard 1015. Analog to digital conversion of radio voice by 2400 bit/second linear predictive coding. National Communication System, Office of Technology and Standards, 1984.

[183] Federal Standard 1016. Analog to digital conversion of radio voice by 4800 bit/second code excited linear predictive coding. National Communication System, Office of Technology and Standards, 1991.

[184] International Standards Organization. Report on the MPEG-4 speech codec verification tests. ISO Publication: ISO/IEC JTC1/SC29/WG11, Oct 1998.

[185] International Standards Organization. MPEG-4 Parametric coding. ISO Publication: ISO/IEC 14496-3 Subpart 2, Mar 1998.

[186] ITU-T Recommendation G.711. Pulse code modulation (PCM) of voice frequencies. ITU Publication, Nov 1988.

[187] ITU-T Recommendation G.723.1. Speech coders: Dual rate speech coder for multimedia communications transmitting at 5.3 and 6.3 kbit/s. ITU Publication, Mar 1996.

[188] ITU-T Recommendation G.726. 40, 32, 24, 16 kbit/s Adaptive Differential Pulse Code Modulation (ADPCM). ITU Publication, Dec 1990.

[189] ITU-T Recommendation G.728. Coding of speech at 16 kbit/s using low-delay code excited linear prediction (LD-CELP). ITU-T Publication, Sep 1992.

[190] ITU-T Recommendation P.861. Objective quality measurement of telephone-band (300–3400 Hz) speech codecs. ITU Publication, Feb 1998.

[191] The International Telegraph and Telephone Consultative Committee. *CCITT Blue Book*. CCITT, Geneva, 1989.

[192] Telecommunications Industry Association. TDMA Cellular/PCS Radio Interface – Enhanced Full-Rate Speech Codec. TIA/EIA/IS-641 Standard, 1996.

Index